YOUR LIFE, UPLOADED

GORDON BELL is a principal researcher at Microsoft. He lives in San Francisco and Sydney, Australia.

JIM GEMMELL is a senior researcher at Microsoft and lives in San Francisco.

YOUR LIFE,

THE DIGITAL WAY TO BETTER MEMORY, HEALTH, AND PRODUCTIVITY

UPLOADED

GORDON BELL AND JIM GEMMELL

FOREWORD BY BILL GATES

PREVIOUSLY PUBLISHED AS *TOTAL RECALL*

A PLUME BOOK

PLUME
Published by Penguin Group (U.S.A) Inc.
375 Hudson Street, New York, New York 10014, U.S.A.
Penguin Group (Canada), 90 Eglinton Avenue East, Suite 700, Toronto, Ontario M4P 2Y3, Canada (a division of Pearson Penguin Canada Inc.); Penguin Books Ltd, 80 Strand, London WC2R 0RL, England; Penguin Ireland, 25 St Stephen's Green, Dublin 2, Ireland (a division of Penguin Books Ltd); Penguin Group (Australia), 250 Camberwell Road, Camberwell, Victoria 3124, Australia (a division of Pearson Australia Group Pty Ltd); Penguin Books India Pvt Ltd, 11 Community Centre, Panchsheel Park, New Delhi – 110 017, India; Penguin Group (NZ), 67 Apollo Drive, Rosedale, North Shore 0632, New Zealand (a division of Pearson New Zealand Ltd); Penguin Books (South Africa) (Pty) Ltd, 24 Sturdee Avenue, Rosebank, Johannesburg 2196, South Africa

Penguin Books Ltd., Registered Offices: 80 Strand, London WC2R 0RL, England

Published by Plume, a member of Penguin Group (USA) Inc. Previously published in a Dutton edition as *Total Recall*.

First Plume Printing, November 2010
10   9   8   7   6   5   4   3   2   1

Grateful acknowledgment is made to the following for permission to reprint previously published material:
pp. 35–36, 38 reprinted from the article appearing in the *Atlantic Monthly*, "As We May Think," Vol. 176, No 1 (1945), 641–649 with permission from Edward Vannevar Bush, grandson of Vannevar Bush.

Thad Starner: excerpts from the CARPE 2006 presentation © 2006 Thad Starner, Georgia Institute of Technology. All rights reserved. Used with permission.

 REGISTERED TRADEMARK—MARCA REGISTRADA

The Library of Congress has catalogued the Dutton edition as follows:

Bell, C. Gordon.
Total recall : how the E-memory revolution will change everything / Gordon Bell and Jim Gemmell ; foreword by Bill Gates.
    p.   cm.
Includes bibliographical references and index.
ISBN 978–0–525–95134–6 (hc.)
ISBN 978-0-452-29656-5 (pbk.)
1. Information society. 2. Memory. 3. Neural networks. 4. Information resources management—Forecasting. 5. Telematics—Forecasting. 6. Computers and civilization—Forecasting. I. Gemmell, Jim, 1965– II. Title. III. Title: How the E-memory revolution will change everything.
HM851.B4525 2009
303.48'34—dc22   2009011526

Printed in the United States of America
Original hardcover design by Daniel Lagin

For Jim Gray

CONTENTS

FOREWORD

BY BILL GATES

Instant, easy access to information has always been one of the most important and exciting promises of the Digital Age. I first spoke about "information at your fingertips" in a speech in 1990. Back then, I described information at your fingertips this way:

> Someone can sit down at their PC and see the information that's important for them. If they want more detail, they ought to just point and click and that detail should come up on the screen for them . . . all the information that someone might be interested in, including information they can't even get today.

It's amazing how far we have come since then. With the Internet, with computers and devices that have orders of magnitude more processing power and storage than we had in 1990, and with search engines and other software for finding and manipulating data and content, our access to information is really quite remarkable. Now we take almost completely for granted the ability to open a Web browser and link to a seemingly limitless number of sources for information on virtually any topic—which is how I was able to find, in just a few seconds, a quote from a speech that I gave almost twenty years ago.

Most people think about this idea of information at your fingers as an improved version of a trip to the library. Of course, it's really a massive global network of linked libraries that contains not just books, journals, reports, newspapers, and magazines, but also information about companies, organizations, products, and services, as well as contributions on any subject you can imagine from experts and nonexperts through blogs and other forms of social communications.

As impressive as this list is, there's something important that's not usually included—personal information and personal experiences.

Every day, we are exposed to astonishing volumes of words, data, and media at work, school, home, in stores, via the Web, on TV—everywhere we go. We interact with many people, some familiar, many we may not see again for a long time, if ever. We have a continuous flow of experiences. There are financial transactions, medical data, school transcripts, family photos—the list goes on and on.

What happens to all of this stuff? We store a small percentage in our brains and file some of it away in paper or electronic form. But the truth is we forget much of it and throw away most of the rest.

It's a lot of stuff to leave behind.

What would happen if we could instantly access all the information we were exposed to throughout our lives? If there were a way to recall everything you once knew about someone you are going to see again for the first time in twenty years? If you could tell your doctor everything you had eaten in the week before you broke out in hives, both yesterday and six months ago?

I can't think of anyone better to begin to find the answers than Gordon Bell. For the last decade, he and Jim Gemmell have been working on a project called MyLifeBits that touches on these very questions.

Now, we've reached the point where these are more than just abstract questions. The capacity is available to store hundreds of hours of video, tens of thousands of photographs, and hundreds of thousands of documents in digital form at a very affordable price. Within a decade, we'll be able to store more than a hundred times this amount of information for even less than it costs now.

Even more important, we're getting closer and closer to having software that will enable you to organize and sort all that information so that it is easy to find what you need, even if you aren't sure what you are looking for.

The MyLifeBits project began as an effort to digitize the books Gordon has written, and it has turned into a pioneering endeavor to record and preserve in digital form everything he sees, hears, learns, and experiences.

The implications of this work are profound and exciting. As Gordon and Jim explain in this important book, the results could very well change the way we think about memory; how we manage our health; the way we share experiences with other people, even other generations; and much more.

It's no surprise to me that Gordon is breaking this new ground. He is one of the true pioneers in computing and it would be almost impossible to overstate the importance of his contributions to progress in this industry, whether it was through his role building the first minicomputers at DEC 1960s and 1970s; his involvement as head of the National Science Foundation's information-superhighway initiative; or the work he has done at Microsoft since 1995 on telepresence and telecomputing.

He is also one of the industry's most important original thinkers, not only about how to advance the state of the art of digital technology, but also about the role that technology plays in society and in people's lives. I have so much admiration and respect for the depth of Gordon's thinking and the quality of his work.

I think I've known Gordon for about twenty-five years. But I'm not quite sure. As I was getting ready to write this foreword, I checked with Gordon to see if he could remember when we first met. I recalled a lobster dinner that Gordon hosted in September of 1983 in Marlboro, Massachusetts, when I first got involved with the Computer History Museum that he founded. He thinks it's possible that we met a few months earlier when he flew to Seattle to discuss licensing DOS for DEC's Rainbow line of computers. Gordon also checked with a friend who was a colleague at DEC at the time, who says it was probably in 1982, when discussions about the Rainbow-DOS deal first got under way.

I wish we could pinpoint our first meeting with a little more certainty. As Total Recall advances, this kind of information will be available to us as a matter of course. I know we'll welcome this change.

AUTHORS' NOTE

Since 1995 the two of us have had a close partnership that has led to the ideas and words in this book. For the sake of readability, we have adopted Gordon's voice for our four hands. *I* is Gordon Bell, and we are explicit when a story is about Jim Gemmell.

We chose not to delve into technical detail in the main text. For those who would like to know more about the computer engineering behind Total Recall, we refer you to the Annotated References and Resources section at the end of this book, and our Web site, www.totalrecallbook.com.

YOUR LIFE, UPLOADED

PART ONE

PART ONE

CHAPTER 1

THE VISION

I'm losing my mind.

Not the Gordon-needs-a-high-priced-psychiatrist kind of losing one's mind, although my teenage granddaughter may disagree. Instead, each day that passes I forget more and remember less. I don't have Alzheimer's or even brain damage. I'm just aging.

Yes, each day I'm losing a little bit more of my mind. By the way, so are you.

What if you could overcome this fate? What if you never had to forget anything, but had complete control over what you remembered—and when?

Soon, you will be able to. You will have the capacity for Total Recall. You will be able to summon up everything you have ever seen, heard, or done. And you will be in total control, able to retrieve as much or as little as you want at any given time.

Right now, if someone had even a single photo from each day of her life, we would be amazed. But soon you will be able to record your entire life digitally. It's possible, affordable, and beneficial.

If you choose, you'll be able to create this digital diary or e-memory continuously as you go about your life. This will be nearly effortless, because you'll have access to an assortment of tiny, unobtrusive cameras, microphones, location trackers, and other sensing devices that can be worn in shirt buttons, pendants, tie clips, lapel pins, brooches, watchbands, bracelet beads, hat brims, eyeglass frames, and earrings. Even more radical sensors will be available to implant inside your body, quantifying your health. Together with various other sensors embedded in the gadgets and tools you use and peppered throughout your environment, your personal sensor network will allow you to record as much or as little as you want of what happens to you and around you.

If you choose, everything you see can be automatically photographed and spirited away into your personal image library within your e-memory. Everything you hear can be saved as digital audio files. Software can allow you to scan your pictures for writing and your audio files for words to come up with searchable text transcripts of your life. If you choose, you can save every e-mail you send and receive and archive every Web page you visit. You can record your location and path through the world. You can record every rise and dip in your heart rate, body temperature, blood sugar, anxiety, arousal, and alertness, and log them into your personal health file.

The coming world of Total Recall will be as dramatic a change in the coming generation as the digital age has been for the present generation. It will change the way we work and learn. It will unleash our creativity and improve our health. It will change our intimate relationships with loved ones both living and dead. It will, I believe, change what it means to be human.

Three streams of technology are coming together to make the world of Total Recall a reality. First, and perhaps most important, we are recording more and more of our lives digitally without even trying. Digital cameras, e-mail, cell phones, and personal digital assistants (PDAs) are the vanguard of technology that is generating an explosion of digital records of our daily lives. Digital sensing and recording will become ubiquitous. Second, this mountain of new personal digital records can now be stored more cheaply than can easily be imagined—for about two hundred dollars you can own enough memory to store everything you read, everything you hear, and ten pictures a day for your whole life. Third, technologies enabling you to search, analyze, and present all kinds of reports from such large mountains of data are being developed, with astonishing results. Google will by no means be the last extraordinarily successful company to be built on new search technologies. So, we live in a world with more digital memories, more space to store them, and better and better technology to recollect them. The world of Total Recall is inevitable for these three reasons.

With the same ease with which you can now search for just about any subject on the Web, you will be able to search your own electronic memory for any arbitrary item of knowledge you have ever encountered, any snippet of conversation to which you have ever been party, any document that has ever passed before your eyes, any place you have ever visited, any person you have ever met. You become the librarian, archivist, cartographer, and curator of your life.

The ability to recover particular events, names, faces, and words is just the most obvious benefit of the Total Recall revolution. Software will allow you to sort and sift through your digital memories to uncover patterns in your life you could never have gleaned with your unaided brain. Your work habits, your leisure habits, and your spending habits; your emotional response patterns in various situations and around certain people; the numerous subtle factors that affect your mental well-being and your physical health; and just about anything else you care to know about yourself can be chronicled, condensed, cross-correlated, and plotted out for you in useful and illuminating ways. Your goals and achievements for time management, budgeting, and balancing all aspects of your life, work, and health can be tracked through progress charts you set up for yourself. Having access to such detailed and personally relevant feedback is one of the most potent spurs to motivation and productivity to be had.

Now imagine a complete digital record of your life, a complete e-memory of your time on earth. Aristotle, Alexander the Great, Shakespeare, Mozart, Edison, and Einstein are dead but their ideas, their deeds, and their personalities have achieved a sort of immortality. Few aspire to be remembered along with history's great characters, but by recording your life digitally you have the opportunity to bequeath your own ideas, deeds, and personality to posterity in a way never before possible. With such a body of information it will be possible to generate a virtual you even after you are dead. Your digital memories, along with the patterns of fossilized personality they contain, may be invested into an avatar (a synthesized persona) that future generations can speak with and get to know. Imagine asking your great-grandfather about what he really loved about your great-grandmother. Your digital self will reach out to touch lives in the future, allowing you to make an impact for generations to come.

The era of Total Recall is dawning, whatever you personally choose to do with the technology. You may embrace full-scale "life-logging," and devote much effort to maximizing your e-memories, or you may prefer to record your activities only modestly and se-lectively, or even reject the whole idea and strive to leave as small a digital life-footprint as possible. People felt and continue to feel disdain for the Internet and even personal computers. Some of us don't want a cell phone. No matter. Whether you are an early adopter, a late adopter, or a never-in-a-million-years nonadopter, society at large is on an inexorable path toward Total Recall technology and it is going to transform the world around you. The power of this transformation will be awesome.

THE E-MEMORY MACHINES

Total Recall is arriving in a blaze of innovation. There are ever more ways being developed and packaged for gathering information from the world and from our lives and putting it to myriad uses. People the world over are gabbing, texting, picture-taking, video-capturing, and Web-surfing on their cell phones. Phones and cameras can now send pictures automatically to a Web site where they can be grouped, culled, and annotated later. Parents take zillions of hours of video footage of their children with pocket-size cameras that dump directly into their home computers. Casual joggers can now analyze their performance at levels once reserved for world-class runners, tracking their metabolic burn rates and the lengths, times, and elevation profiles of their runs using small, affordable devices worn on their bodies. You can buy a bathroom scale that automatically sends your weight to an encrypted Web site where you can examine your progress (or lack thereof) in cold, hard, objective numbers. College students can time-sync their typed notes to their audio recording of a lecture, allowing them to relisten to part of the lecture later by clicking on one of their notes.

This cornucopia of information-gathering devices continues to grow in size and diversity while the devices themselves continue to grow smaller, cheaper, and more multifunctional. Meanwhile the cost of digital memory continues its exponential descent. When it comes to recording information, the technology stream is gushing toward ubiquity and saturation, toward a world in which price and convenience will no longer be factors when deciding what or whether to record. Indeed, we are headed toward a world where it will require a conscious decision (or a legal requirement) not to record a certain kind of information in a certain time or place—the exact reverse of how things are now. The technological and economic forces driving this trend are strong. Arguably, only a vast legal or political effort of social engineering can prevent it from effecting far-reaching changes in the way modern life is lived. That sort of catastrophic counterrevolution sounds far-fetched, but there are more realistic scenarios that I will discuss in Chapter 8.

E-memories will provide every person who embraces them with a different sense of their whole lives. It won't erase human nature's capacity for self-deception, but it will surely make the truth of what we did and what happened around us more available, clearer, and less obscured by nostalgic make-believe. The benefits will also be distinctly practical. In the chapters ahead I will describe these benefits in the workplace, to our health, and in our capacity for learning. Higher productivity, more vitality and longer life spans, deeper and wider knowledge of our world and ways to accomplish things in it—these are all wonderful practical consequences of this coming technological revolution. But there will also be psychological implications. Enhanced self-insight, the ability to relive one's own life story in Proustian detail, the freedom to memorize less and think creatively more, and even a measure of earthly immortality by being cyberized—these are all potentially transformational psychological phenomena.

But is it really feasible to record everything that happens in a person's life? Shockingly perhaps, the memory needed to store a person's lifetime of recorded experience is already here and affordable and is always growing cheaper. The rate of price decrease is given by Moore's Law, which states that the transistor density that can be etched onto the silicon wafer of a microchip doubles every two years. This means that every two years, the cost of computer memory is cut in half—so you can afford to buy twice as much as last year. Moore's Law was first published in 1965 and has held up with remarkable consistency ever since.

The growth of digital storage capacity has been staggering. In 1970, a disk that could store twenty megabytes (twenty million bytes) was the size of a washing machine and cost twenty thousand dollars. Today a terabyte (one trillion bytes) costs a hundred dollars and is the size of a paperback book. By 2020 a terabyte will cost the same as a good cup of coffee and will probably be in your cell phone. One hundred dollars will then buy you around 250 terabytes of storage, enough to hold tens of thousands of hours of video and tens of millions of photographs. This should satisfy most lifeloggers' recording needs for an entire life.

In fact, digital storage capacity is increasing faster than our ability to pull information back out. Once upon a time, you had to be extremely judicious and stingy about which pieces of data you hung on to. You had to be thrifty with your electronic pieces of information, or bits, as we call them. But starting around 2000 it became trivial and cheap to sock away tremendous piles of data. The hard part is no longer deciding what to hold on to, but how to efficiently organize it, sort it, access it, and find patterns and meaning in it. This is a primary challenge for the engineers developing the software that will fully unleash the power of Total Recall.

Where is the desktop PC in this Total Recall revolution? Its impending downfall has been predicted by Silicon Valley denizens for years. I believe the PC is destined for a demotion but is unlikely to vanish. The *P* in *PC* will still stand for "personal"—in fact, it will be more personal than ever before, involved in every aspect of your life. But the *C* will change from *computer* to *computer ecosystem*. Your desktop computer will be just one of many tools at your disposal for e-memory management. You will own an assortment of small, fungible, more modest computers—in fact, you probably already do. They are in your cell phone, in your appliances, and in your car. Increasingly, they will be in your clothes and on your body. They will be virtually everywhere.

C will also stand for *cloud*. This refers to a new way of using the Internet in which data gets stored and software is run "out there" in the abstract ether of cyberspace—in the cloud—rather than locally on your own PC. Ultimately, the cloud turns information processing and storage into a metered utility just like electricity and water (with the difference that there will be many free offerings). It's like having the power of computers "on tap."

With cloud computing, your data becomes untethered from particular devices. Your e-memory follows you wherever you go, accessible from any device you happen to be using. You, not your desktop's hard drive, are the hub of your digital belongings. Many of us experience this now with our e-mail, which we access from our desktop PC, our notebook PC, our smartphones, or any device we might borrow with a Web browser. Increasingly, everything in your e-memory will be accessible anywhere, anytime, from any device at hand.

Of course, many of your devices will have vast storage; your cell phone will hold more than a terabyte, and your notebook computer will carry more than two hundred terabytes. So think of every device you own as part of the cloud too. You can tap into a service provider, but you can also tap your home servers, your portable devices, and possibly those of your friends (they may keep a backup copy for you). And chances are that a copy of most things you will want will already be there in your notebook or cell phone.

Just as the cloud will offer digital storage, it will also offer incredible processing power. This will range from your cell phone asking for extra help from a couple of machines in your home, to paying a fee to rent a few thousand computers from some service provider for a couple of minutes. Your health information might be mined for patterns by a couple of servers in your home. A service provider might keep the index of all your information up to date so that your cell phone isn't slowed down by index "crawling."

Of course, even today's smartphones already boast impressive processing power, doing things like voice recognition, movie playback, and running full-blown databases. For most tasks you will already have enough computing at hand. However, cloud processing isn't always just about computational muscle. Cloud processing can also provide simplicity. Instead of installing and maintaining software in all of the computing devices in your home, it can be much simpler to install to just one home server, and have the other devices just act as terminals to the server. Then you only have one computer that requires upgrades, license verification, and other management drudgery. The same can happen across the Internet, and we already see Internet-based alternatives to traditional desktop software, including e-mail, instant messaging, word processing, spreadsheets, and backups.

In the future, you probably won't know or care whether an application is running on your device, or whether it is running in a cloud server and your device is just a terminal. Ray Ozzie, Microsoft's chief software architect, describes it this way:

> All your devices will be appliance-like. You'll buy them
> to suit your need or tastes. Phones, PCs, Xboxes, whatever.

You'll go to the Web and license software that is intended to be used on these various appliance-like devices. When you turn it on and log-in for the first time and "claim" the device as yours, the things that should be on it (the right data, the appropriate version of a given app for that kind of device) simply appear. When you run the app and use the data, it is automatically sync'ed so that there's only ever temporarily a "dirty" copy of data or settings on the device that isn't also present in the cloud. When you recycle a device and "unclaim" or "disown" it, your stuff vanishes.

Many Web service providers are in the same boat with you. They often don't care where your data or where their own data actually resides, because they, too, may outsource their storage and processing needs, paying for whatever capacity they use when and as they use it. Many companies run their entire businesses using remote servers, without having to invest in computers, storage, or associated software. To serve this need, Amazon, which has vast amounts of excess storage, has a service called EC2, for Elastic Compute Cloud. Likewise, Microsoft has Azure, a cloud-based operating system that will let companies develop and run Web applications without setting up their own data center. Microsoft CEO Steve Ballmer has predicted that nearly all Internet data centers will be outsourced in this way by 2020.

Cloud computing will lead to a single, integrated e-memory experience. Every device will act as an access point to recall from your e-memory. And every device will also become a source of information feeding into your e-memory, helping to record your experience.

Most people's cloud-interface device of choice is going to be a small, lightweight device that combines the functions of a cell phone, a camera, a personal digital assistant, a Web browser, an MP3 player, a GPS locator, and any other sensors and functions that can be crammed into it. Early versions of these devices are already abundant. They're called smartphones (e.g., iPhone, Black-Berry, Windows Mobile, Symbian, Palm, and Android). As smart-phones become better at many traditionally PC-based functions, trusty but less agile laptops are being left at home. Sales represen-tatives tap into customer databases five minutes before meeting with clients. Managers track inventories in real time. Physicians call up medical records and lab reports while standing at a patient's bedside.

Your smartphone plus whatever sensors and miscellaneous devices you wear and carry will all be linked together to form a personal digital memory collection-and-management system that will (if you choose) be able to record just about everything you see, hear, and do and keep it all in one big virtual collection in the cloud. The uses of such an archive are limitless.

Consider the fact that currently, nearly every financial trans-action you make—making a deposit, withdrawing cash, pay-ing with a credit or debit card—is registered electronically as a unique event. Every month when you open your statements you can see the trail you have left, which is geographical as well as financial, with entries like "01/07/08 32.60 **Australian Dollars 29.10 MOS CAFE SYDNEY NSW; 01/08/08 VIZZVOX INC WY 99.00; 01/12/08 MICROSOFT *ZUNE OFFICE SUPPLY STORE 00877-438-9863 WA 15.00"; and so on.

Imagine extending this trend to all the recordable events in your life that you can imagine ever wanting to be able to recall or examine or contextualize at a later date: where you went, how you got there, who you met, what you did, what was said, how your vital signs varied, who you called, what you read, what you wrote, what you looked at, what pictures you took . . . all these things and more can be automatically recorded and saved, indexed, filed, and cross-referenced by time, location, and other natural linkages to make them easy to refind later and to sift through for patterns and trends.

While much of the technology for Total Recall is already available—e-mail, cell phone, camera, home videos, social-networking sites, photo- and video-management sites, and so on—these many pieces remain isolated and fragmented. They are not yet integrated by a single set of tools or unified under a common interface. The current e-memory ecosystem is relatively fraught with inconveniences: nonportable data formats, the need to keep on top of dozens of passwords and personal profiles, short battery life, and data fragmented across devices and applications. As these impediments disappear, led by the shift to cloud computing and the evolution of hardware, the e-memory experience will be transformed, and the technology of Total Recall will become a reality in most people's lives.

FEAR

Such a massive change can be frightening. Won't the government access all our e-memories and spy on us? Since George Orwell published his masterpiece novel *1984*, the idea of government as Big Brother has loomed large in rational critiques of real government policies as well as in conspiracy theorist outcry. We like to use the term Little Brother. If Big Brother rules the authoritarian vision of a surveillance society, Little Brother rules the "democratized" vision. It is a society of omnipresent surveillance in which the recording equipment is not controlled by a single central authority, but by millions of individuals and private entities. Total Recall is perfectly consistent with social values behind the inspiration of the Internet, in which I'm proud to say I also played an early role.

Our culture will need to develop a whole new body of etiquette about who may record whom when and where. Our sense of privacy will continue to evolve, as it has since everyone knew everyone else's business in the village life of agricultural economies. There is more to say on this subject and potential unintended consequences in the coming age of Total Recall. I'll get to those issues in the third part of this book, after we look at the big impacts it will have on these key areas of life: the workplace, our health, our capacity for learning, and our most personal intimate relationships. In the remainder of this first part of the book we'll look at a bit of the history that got us here, my part in the revolution so far, and what the proliferation of e-memories and their use is going to do to the memories in our heads.

HOW DID WE GET HERE?

The arc of human development from the Stone Age through the present can be seen as an ongoing quest for Total Recall. One thing that has defined our progress as the preeminent species on the planet has been our ability to develop better and better systems of memory.

Our greatest innovation was language, a unique system for representing, storing, and sharing knowledge. Language made us into the first and only truly cultural animal, able to share both highly specific and powerfully abstract bits of knowledge across societies and down through generations.

The next great turning point in human development was the invention of writing, which it became necessary to invent as the needs of record keeping in agrarian city-states outstripped the limits of naked memory. Thanks to writing, human knowledge snow-balled over just a few thousand years and brought us most recently into the Information Age. Around the middle of the last century the digital computer joined our mnemonic arsenal and rapidly pre-cipitated another epochal change in how we manage our knowl-edge. A mere generation ago, the amassing of information was so expensive that a world of Total Recall could be no more than a wild science-fiction dream.

But themes of Total Recall have been explored in science fic-tion for decades.

In *Hominids,* Robert J. Sawyer imagines the citizen of the future sporting a body-implanted "companion" computer that transmits information about his or her location, as well as three-dimensional images of exactly what he or she is doing, to an "alibi archive." The archive protects against false accusations.

In *The Truth Machine,* James Halperin describes a world where not only is recording common, but everyone testifies using a per-fect lie detector. Crime is drastically reduced. But when people want to talk candidly in a free and open discussion, they must turn off their recording equipment.

The 2004 movie *The Final Cut* depicts a world where people pay to have their babies' brains implanted with memory chips, called Zoes, that record everything those children see and hear through-out their lives. When the person dies, the chip is removed and a professional "cutter"—in this case a somber Robin Williams—goes through the chip's footage to edit his or her life down to a (flatter-ing) feature-length movie called a Rememory, which is played for friends and family at the memorial. Cutters can make "saints out of criminals," as Williams's character does with the life of a child abuser. The movie also shows protesters with placards demanding "the right to forget" and darkly depicts the lengths to which some people might be willing to go to get their hands on the private life recordings of a political enemy.

Another common theme of sci fi is digital immortality, whereby a person's lifetime of experience, knowledge, and personality are simulated by a computer. In the Superman movies, the Fortress of Solitude can create an on-screen likeness of Superman's wise and stately father, Jor-El, which is able to answer questions about Kryptonian history, technology, and culture. In the British TV sitcom *Red Dwarf* the last human in the universe, David Lister, is forced to endure the company of a hologramatic simulation of his insufferable prat of an ex-crewmate, Arnold Rimmer. And in the American TV espionage series *La Femme Nikita,* the character Madeline is virtually resurrected in a similar fashion. As the character Quinn explains: "Madeline's psychological and analytical profiles were extensively documented. It was a question of merging them with [an] artificial intelligence program." Thus the steely intelligence director—or at least, the benefit of her lifetime of experience— continues to aid the living after her death.

Science fiction can be fun and stimulating, but one of the best recent sketches of what Total Recall might actually look like comes from Donald Norman's 1992 nonfiction book, *Turn Signals Are the Facial Expression of Automobiles.* Norman, an expert on human-machine interface and design, proposes that in the future everyone will have a lifelong companion he called the Teddy—a "personal life recorder."

In Norman's vision, this device would be issued very early in life, perhaps at age two or three, and dressed up in the guise of a toddler-friendly toy. Because the devices would be in the hands of young children, the first life recorder would be soft and cuddly, like a plush bear—hence the name Teddy. The Teddy needn't be limited to just the passive recording of a child's actions and words.

It could be designed to be interactive, and help him or her learn to read, write, draw, and sing.

By starting so young, the Teddy would end up storing a great deal of your life experience. You would become quite intimate with your Teddy. It would "know" all about you and could answer questions about your past. At the same time, it could give you access to knowledge and information from the Internet and other sources. When you outgrew your stuffed-animal phase, your Teddy would change form to match your growing sophistication and interests. Its guise would change, but its complete record of your personal experiences and knowledge would always follow you.

THE KIDS ARE ALL RIGHT

The Teddy is science fiction, but just barely. Total Recall is coming, and young children might be one of the first segments of society to have it applied to them. Many parents won't be able to resist. Today's children are already extensively surveilled. They are monitored by microphone while they sleep. They are relentlessly photographed and videoed by camera-happy grown-ups. Some are even fitted with directional radio or GPS tracking devices while out and about. Some preschools provide live Web cameras so parents can look in on the kids' activities from work. Some parents buy "nanny cams," which allow them to spy on their children's caretakers and other domestic help. (Superficially echoing Norman, nanny cams can be bought in the guise of teddy bears.) Older kids are given cell phones so that they, unlike every previous generation, never have any excuse for being unreachable or untraceable for longer than the time it takes to drive through a tunnel.

Parents do all this for two reasons: for their children's safety (or perceived safety), and to create a trove of memories of what

they were like at each fleeting stage of development. The new
wave of cheap and unobtrusive recording devices extends parents'
already widely exercised ability to monitor and record their precious charges. Given today's ethos of "hyperparenting," it's hard to
imagine the trend stalling or reversing anytime soon. For all these
reasons, we can expect one of the vanguards of lifelogging to be
children. These are the same future citizens who will be the most
enthusiastic and least conflicted about embracing the technology
of Total Recall.

But we don't have to wait for the current crop of toddlers to
grow up to see if this prediction will come true. The current crop
of young adults is itself a convincing case for the inevitability of
Total Recall as common practice.

The Millennials, also known as Generation Y, are the cohort of
Americans born approximately between 1982 and 2001. They came
of age with Google, cell-phone cameras, file sharing, text messaging, Facebook, MySpace, YouTube, Second Life, and Twitter (the
online-community service where friends and families post fre-
quent, 140-characters-or-less "microblog" entries about whatever
they happen to be doing or thinking at that moment). A few formed
software companies and became millionaires in their twenties.

The seventy-million-plus Millennials are adept at multitask-
ing. They listen to music, do homework, watch TV, and send instant messages—simultaneously. Nearly all own cell phones and
computers. They snap pictures wherever they go. They social-
ize quite a lot online, chatting, trading music, and playing video
games. Many stay in touch with their cliques so obsessively, they
can't bear to turn off their cell phones for a fifty-minute classroom
session. If they misplace their smartphone, they feel as if they've
lost their minds.

Of course, no group of millions of people is going to be monolithic. Studies show that not all Millennials are as tech-savvy as legend would have it, but even the unsavvy ones tend to be much less intimidated by technology they don't understand than their parents would be. This generation makes far less distinction between their private and public lives than even slightly older generations. The Internet is littered with millions upon millions of their blog entries, their photo streams, their fan fiction, their chat-forum entries, their comments on all of these things, and of course their comments on comments. Scads of them post everything from their youthful hijinks to their intimate confessions on YouTube. Posted videos often elicit avalanches of video responses, many of them raw and unrehearsed, showing the responders' unfiltered, unedited reactions—on a site that literally billions of people can view.

Those who put their lives up on the Web for others to view are called life bloggers (*blog* being short for "Web log"). I am a lifelogger, not a life blogger. That is, I log my life into my e-memory. I may be old-fashioned, but it strikes me as foolish to publish too much, especially to an unrestricted audience. There's too much risk and too little benefit. My lifelogging is personal and private. I do it for the very pragmatic value that it gives back to me. Unlike those making the effort to create blog entries and YouTube clips, most of my lifelogging is automatic. When I share, I do it cautiously, considering the trustworthiness of the individual recipients. Public publishing is only for what I am glad to have the world associate with me—forever. Once out on the Web it is easily copied, so you cannot "take it back"—it has become part of the permanent cyber landscape (or landfill) forever. If you are one of those who really want to share everything with the world, go ahead, it is your right. But I wouldn't recommend it to anyone.

While the first decade of the twenty-first century has seen a proliferation of digital life-chronicling, much of it is still ephemera, tossed-off throwaway documents like nineteenth-century theater billings and political pamphlets (which historians now drool over, by the way). Most of this chronicling is still haphazard: Parties, outings, and weekends are extensively digitally commemorated, but drives to school, study sessions, and dinner at the grandparents' are not. I've observed people who are happy to record minutely from their lives, but they aren't systematic about it. They record the things they think are important or cool or memorable at the time, but they don't yet record everything they see, hear, do, send, receive, read, and compose.

This isn't necessarily reticence on their part, because the software tools and hardware accompaniments for easy lifelogging aren't readily available yet—not quite yet. Through the decade of the 2010s, user-friendly lifelogging applications will proliferate just like any other software niche, and a plethora of cheap devices for sensing, tracking, and compiling all kinds of information from all corners of life will pour steadily into the consumer-electronics market. As this happens, we will see lifelogging start to catch on with the Millennial generation, and older generations too.

COLLECTING E-MEMORIES, DISCOVERING WHO YOU ARE

It's impossible to know exactly how long it will take for lifelogging to become common practice, but it's almost a sure bet that it will do so within a decade. Abstaining from lifelogging will begin to seem more like avoiding the use of e-mail or cell phones, because so many advantages and conveniences will be foregone. Those who shun recording will be less empowered than those who embrace it.

You probably already spend a good deal of time each year filing away receipts, checkbooks, financial statements, photos, article clippings, and sentimental or souvenir items such as birthday cards and ticket stubs. You probably also take some time to label and annotate certain items to make them easier to later refind them and figure out what you kept them for.

Total Recall just means storing and annotating things digitally instead of physically. It will not be any more time-intensive; in fact it will probably be less time-intensive, and the amount of information will be orders of magnitude larger. Digital records take less time to file, take up almost no space, and are easy to search. Pioneers like me might be manually filing records by scanning or snapping photos of them, typing or speaking quick notes about items that need explanation, even composing longer stories as I create the record. But soon so much of that will be as automated as your bank account statement.

You won't have to worry about forgetting someone's name, face, or details of the conversations you have throughout each day. When you want to recall what someone said, you'll be able to search for phrases or keywords. If the search brings up too many results, you'll be able to narrow it down by other criteria: I remember it was said while I was on a trip to Atlanta. The person who said it was a woman, and I think she wore glasses. I'm sure it happened before I took my current job. Given enough criteria, even vague ones, a good search program will usually be able to find exactly what you are looking for.

Imagine the ability to scan the past with the ease that would put Google to shame. Imagine how it could affect therapy sessions, friendly wagers, court testimony, lovers' spats (of course, metajudgments like "It's the way he said it" or "You didn't really mean it" will never go away). Imagine how easy it will be to prove that repairs were done, that a salesman went back on his word, or that the dog really did eat your homework. Think of how nice it would be to have recordings of childhood conversations with your best friend, or a complete audio library of the millions of priceless things your kids said when they were toddlers. What were those first baby words, really?

Just as important as the ability to search will be the ability to data-mine your e-memory archive, to find correlations and multi-dimensional patterns in your life experience. Your e-memory archive could give you insight into how you spend your time. Click a button and see a chart of how much exercise you have been doing in the last month, or year. Compare it to what you did when you were sixteen, or in the summer versus the winter. Or check how often you smile. Compare that to before you were married—or divorced. Total Recall can be a time-management gold mine, allowing you to define your goals or set standards for yourself and then track how they compare with your actual behavior. Maybe you are spending too much time managing your e-memories. Check it.

With the right software you will be able to mine your digital memory archive for patterns and trends that you could never uncover on your own—graphing, charting, sorting, cross-sectioning, and testing for hidden correlations among all your bits. Imagine if you could bring into a single database all the pictures you take, all the places you visit, all the routes you take, all your notes and annotations, all your e-mails, along with room temperatures, weather conditions, diet, activity, whom you met with, your meeting cancellations, what you read, when you worked, what TV shows you watched, your mood swings, your flashes of inspiration. What would happen if you could take that whole slurry of life-history fragments and run it all through a powerful pattern-detection program? What kinds of patterns might you find?

Digital memories can improve your health and extend your life. Equipped with new generations of personal sensing devices, you'll be able to collect torrents of physiological data from yourself—alpha waves, dips in cortisol, temperature, pulse, sweating, and scores of other measures—in real time.

And the health benefits of Total Recall aren't limited to those with known health risks. Whether it's sticking to your exercise plan, watching your weight, battling insomnia, managing allergies, tracking down the causes of a recurring rash, gaining control of your stress and anxiety responses, or training your mind to focus better, your best and most often used tool is going to be an easily and totally recallable continuous physiological e-memory.

TOTAL RECALL

I hate to lose my memories. I want Total Recall.

This isn't a pipe dream. I know that three streams of technology advancement—recording, storage, and sophisticated recall—have already launched the beginning of the Total Recall era. It is absolutely clear that by 2020 these streams of technology will have matured to give the complete Total Recall experience.

I don't work on anything unless I see a practical payoff. I got started in this by wanting to get rid of all the paper in my life. Then I wanted better recall; then a better story to leave to my grandchildren. Soon I became aware of potential benefits for my health, my studies, and even a sense of psychological well-being from decluttering both my physical space and my brain. As time passes, I become more and more excited about the benefits of Total Recall.

As you read on about Total Recall, I'll tell you more about my

own story, and I'll elaborate on the incredible gains that Total Recall will supply across so many areas of life. In the last section of the
book, I'll discuss how to put these ideas into practice, and explain
how you can stop losing your mind and get started creating your
own e-memories.

CHAPTER 2

MYLIFEBITS

My own quest for Total Recall began in 1998 while I was working as a researcher at Microsoft. I didn't start out thinking of Total Recall. As usual, I was being pragmatic and looking for things to make my own life better. A colleague, Raj Reddy, asked me if he could digitize the books I had written and put them on the Web as part of his Million Books Project.

"Sure," I told him, "Microsoft has a lot of lawyers. They should be able to get me out of any trouble that comes from copyrights."

Seeing those books become digital felt good, and encouraged me to try some more scanning. I did the scanning myself just to see how to do it and whether it was interesting and useful. I scanned a pile of correspondence, patents, and around a hundred articles. I became even more upbeat, and set my sights on scanning all of my papers and notebooks. I saw it would declutter my office, and allow me to work from home or anywhere I happened to be. I could be an efficient, paperless teleworker.

Then I thought: Why stop there? With all those cheap terabytes of storage coming down the pipe, why not just keep *everything*? Not only books and papers and e-mails, but slide presentations, product brochures, health records, interviews, photos, songs, movies—all the information of my life.

It wasn't as if no one had thought of this before. Bill Gates wrote in his 1995 book, *The Road Ahead,* "Someday we'll be able to record everything we see and hear." Clearly, that someday will come because we'll be able to make practical use of all those e-memories. That someday is going to be in the middle of someone's life. Why not mine? Why not now? But how? How could one person speed the arrival of the era of Total Recall? I became intrigued with the idea of keeping everything.

There is a strong social prejudice against this very simple idea. Keeping everything is like the eighth deadly sin. You'll become a packrat, a horder, obsessed with your past. You shouldn't look back. You need to clear out your attic and throw stuff away. And in a nondigital world, that kind of thinking made some sense. But in a digital world, with time and cost barriers melting away before our eyes, things have changed. Keeping everything doesn't mean you have to spend all your time looking after masses of paper and stuff. Don't throw it away, digitize it.

I especially wanted to rid myself of my filing cabinets and the countless banker boxes holding my old papers. Making them e-memories gave me the pleasure of getting rid of them—without really getting rid of them.

The idea is simple to state: "Record everything, keep everything." But actually putting it into practice turned out to be a major project. Even though cheap terabytes were still some years off, I felt it was important to start immediately. When the day of the cheap terabyte arrived, I hoped to be able to give people insight into the

logistics, costs, benefits, feasibility, and desirability of recording everything. And just what "everything" in your life might mean.

Building my own e-memory became a three-pronged effort.
First, I had to make digital copies of everything from my past. Second, I had to start recording and storing everything I saw, heard,
and did from that point forward. Then, third, I had to figure out
how to organize the information in my digital corpus. This last
prong was crucial. Just saving files willy-nilly into an e-memory is
easy, just as throwing receipts into a drawer is easy; but come tax
time, or if you ever need to find a specific subset of those receipts,
you'll rue your lack of filing discipline. So the big task would be to
figure out what kind of software would be needed to make such a
massive and miscellaneous collection of information useful.

By January 2001 my sixteen-gigabyte e-memory contained
more than five thousand photographs and about one hundred
thousand pages of paper: letters, memos, bills, receipts, financial
statements, legal documents, ticket stubs, business cards, greet-
ing cards, brochures, meeting agendas, symposium programs, di-
plomas, warrantees, manuals, purchase orders, circuit diagrams,
employee evaluations, annual reports, newspaper clippings, article printouts, stock certificates, report cards, childhood drawings,
birth certificates.

I'd hung on to those hundreds of pounds of yellowing paper
not because I wanted to help found a thriving community of sil-
verfish in my home, but because I knew that someday, for some
reason, I would certainly need to refind at least one old item. The
vast majority of them would never see the light of day again, but I
had no way to predict which one I'd need back. I couldn't possibly
know in advance which check I might need to end a payment dispute or whether I would face a tax audit for a certain year. So I'd
felt trapped into keeping all of them. It took me more than a decade

to throw away circuit-theory class notes from MIT, even though it was clear I wouldn't be designing any of those kinds of circuits.

Scanning and digitizing that much paper turned out to be a very big job, so in April 1999, I hired Vicki Rozycki as my personal assistant. Over the next two years we would scan, a handful at a time, what we then thought was everything. Then, for several years after that, we continued unearthing more to scan. It took up a large amount of her time. The hard part was finding stuff and getting it ready for the scanner. (Nowadays there are commercial services that will do this sort of thing for you much faster and cheaper, using automated bulk scanners.)

I never knew quite how much I'd resented the need to stockpile so much paper until I saw it dwindle away like dirty old winter snow in the spring thaw. Folder by folder, box by box, week after week, it disappeared. The clutter and hassle of keeping paper files had been like the half-noticed droning of an electric motor that suddenly goes silent, leaving me in a startled state of peace. Not to make light of tragedy, but this passage from a recent novel, *Extremely Loud and Incredibly Close* by Jonathan Safran Foer, struck a chord with me:

> [It] was the paper that kept the [World Trade Center] towers burning. All of those notepads, and Xeroxes, and printed e-mails, and photographs of kids, and books, and dollar bills in wallets, and documents in files . . . all of them were fuel. Maybe if we lived in a paperless society, which lots of scientists say we'll probably live in one day soon, Dad would still be alive.

I also made digital records of all my physical memorabilia. The scanner worked fine for smallish, flattish items such as medals, coins, and pins, but for larger and more fully three-dimensional objects I had to use a digital camera. I took down all my paintings and made high-quality photographs of them. One of my favorites, titled *Meeting on Gauguin's Beach,* began as a sketch rendered in 1988 by a computer program called Aaron. Then the artist who developed Aaron, Harold Cohen, hand-painted a version for me in vivid oil paints on a five-by-seven-foot canvas. Now I completed the cycle and sent it back to cyberspace.

I took pictures of all my eagles. I love eagles and have amassed a collection of eagle sculptures, picture books, postcards, knick-knacks, and hand puppets. I took pictures of mugs. I have a collection of special coffee mugs. A few of them have eagles, but all of them have some connection to my past. I call some of them my "one-hundred-thousand-dollar mugs" because they are all I have to show for a one-hundred-thousand-dollar investment in a start-up company. I photographed all of those along with their companion T-shirts. The most rewarding part was putting them all in a box and delivering the whole collection to the Computer History Museum—they were someone else's clutter now!

If any of these treasures are ever lost in an earthquake or a fire, I'll have nice remembrances of them. And if my heirs don't want to hang on to the cluttered remnants of my life after I'm gone, they'll always have these images and my notes about them if they want to see what was important to me.

Of course, my collection of two hundred CDs also needed to be ripped onto my computer. I also had several drawers- and shelves-ful of home movies, videotaped lectures, and voice recordings on audiocassette that were collecting dust and needed to be digitized. A service converted the 8mm movies to VHS tapes, which were later converted to CDs.

The second prong of getting my life digitized was to start scanning and recording everything I did from that point forward. In 2002 I decided to go paperless and to scan and not store or file any paper documents. I had already resolved in 1995 not to take any more paper newspapers. (Later, I was at a dinner with *New York Times* publisher Arthur Sulzberger Jr., and his description of their investments in new printing plants convinced me I was right in believing that paper newspapers were on the wane.)

For new paper it was easy. When billing statements or important notices came in, I scanned them in less time than it used to take me to physically file them away. And thankfully, the amount of paper I've needed to scan has shrunk each year since I began. If you piled up all the paper mail I used to receive annually, it would measure about thirteen feet high. Twelve of those feet were utter junk that could be safely thrown away—generic credit card offers, random bulk catalogs, "you may already be a winner" mailings, and the like. (Note: My goal is to record everything I actually read, not what others send me. It's my choice, not theirs, that counts.) Only the remaining foot of paper was worth scanning, and even that has grown less as I've switched to digital bills and statements.

Nowadays it's much easier to avoid paper. All the technical magazines and news sources I read are "born digital," as are nearly all books. For legal reasons I keep a few items in paper, such as stock certificates, but not many. At that time, I started signing everything I could digitally, avoiding the creation, transmission, and especially storage of any paper.

As scanned documents and pictures piled up into my surrogate memory, I faced the challenge of figuring out how best to organize it. I began with what I had to work with: the folder hierarchy that every computer user is familiar with (every folder contains a list of files and subfolders, and every subfolder contains its own list of files and sub-subfolders, and so on). I filed my documents in folders according to a set of categories that made sense to me. From the design point of view this was not perfect, but I had to get started somehow.

In this earliest stab at organizing my scanned data, I split my e-memory into two top-level divisions: items related to current events in my life, and an archive of older, inactive information. Under those two main folders, I had dozens of subfolders for categories including books, medical records, the Computer History Museum (which I'd helped start), trips, underwater photos, food, and so on. Under "Animals" you could find a picture of alligators, various images of San Francisco's wild parrots, and an astonishing set of images showing a snake swallowing a kangaroo whole. I had my "Eagles" folder, of all things eagle-related, and a "Fun" folder, which included a picture of the adult me swinging from a rope.

To help find things again easily, I gave each item a long, detailed file name. For example, the file name of a technical article would include the title, where it was published, the date, keywords, and other pertinent details.

But even with all my documents and pictures stashed away in a well-thought-out classification hierarchy of file folders, it was hard to find particular items quickly, if at all, because it required remembering where it was put. It was just like a library organized by subject without a card catalog. Poring through multiple folders for the right name or thumbnail icon took too much time. Without better labels, even my photos were not much use. When I looked at some, I couldn't recall what they were about. It was painfully clear that the problem would get far, far worse once I started adding hundreds of daily pictures and hours of daily audio to the jumble.

My friend and boss, Jim Gray, teased me about it. When you burn data onto most compact discs, the operation is permanent, and this is known as "Write Once, Read Many" or WORM. Jim mocked me as the inventor of WORN: "Write Once, Read Never."

"It's all just a bunch of bits unless it's annotated," I grumbled.

I began to realize the magnitude of what was lacking. This was not a project to store my life bits; it was about how to get them back!

Scanned documents are image files, not text files, and as such, they're invisible to keyword searches. But with thousands upon thousands of documents in my e-memory, keyword searching would be the only way to re-locate an old file that I could only recollect one or two fragments of, such as a name, a dollar amount, or a dateline. So I ran all the scanned documents through optical character recognition (OCR) software, which is able to recognize written letters and numbers in an image and reconstruct them in a text file. What I ended up with were thousands upon thousands of text files that were neatly interleaved among the scanned files.

Now I just needed desktop search software, that is, software that would allow me to search through my thousands of files for some desired text, just like you search for Web pages now using Yahoo or Google. But at this time operating systems were still several years away from offering desktop search. Desktop search was in its infancy, and every such product I tried was pretty "bleeding edge."

I tried to get Microsoft to take the lead in desktop search, starting with the acquisition of a leading start-up, but was unable to convince the right people. I would have to wait for others to revolutionize search technology. In the meantime, if I wanted to continue my little lifelogging experiment, I would have to cobble together my own solution. In October 2001, Jim Gemmell and Roger Lueder, who had been working with me on other projects at Microsoft, decided that this would make a great research project for them to get involved in. We started out like we do with any new research project, by combing through the published literature to see what others had learned.

I dug up an old paper that I recalled as being relevant, and was surprised at just how relevant it was. In fact, it specified a system almost made to order for us. That's pretty amazing, when you consider that it had been written more than fifty years earlier.

MEMEX

In 1945, when electronic computers were actually multistory buildings, the director of the federal Office of Scientific Research and Development, Dr. Vannevar Bush, published an essay in the *Atlantic Monthly* titled "As We May Think," which outlined a radical new vision of how people might one day keep their own libraries of personal media. He proposed the memex:

> A memex is a device in which an individual stores all his books, records, and communications, and which is mechanized so that it may be consulted with exceeding speed and flexibility. It is an enlarged intimate supplement to his memory.
>
> It consists of a desk, and while it can presumably be operated from a distance, it is primarily the piece of furniture at which he works. On the top are slanting translucent screens, on which material can be projected for convenient reading. There is a keyboard, and sets of buttons and levers. Otherwise it looks like an ordinary desk.
>
> Most of the memex contents are purchased on microfilm ready for insertion. Books of all sorts, pictures, current periodicals, newspapers, are thus obtained and dropped into place. Business correspondence takes the same path. And

there is provision for direct entry. On the top of the memex is a transparent platen. On this are placed longhand notes, photographs, memoranda, all sorts of things. When one is in place, the depression of a lever causes it to be photographed. . . .

. . . As he ponders over his notes in the evening, he again talks his comments into the record. . . . He can add marginal notes and comments . . . and it could even be arranged so that he can do this by a stylus scheme. . . .

Another way to get material into the memex was with a wearable camera:

The camera hound of the future wears on his forehead a lump a little larger than a walnut. It takes pictures. . . . The lens is of universal focus. . . . There is a built-in photocell on the walnut . . . which automatically adjusts exposure for a wide range of illumination. . . . It produces its result in full color. It may well be stereoscopic, and record with two spaced glass eyes. . . .

The cord which trips its shutter may reach down a man's sleeve within easy reach of his fingers. A quick squeeze, and the picture is taken. On a pair of ordinary glasses is a square of fine lines near the top of one lens, where it is out of the way of ordinary vision. When an object appears in that square, it is lined up for its picture. As the scientist of the future moves about the laboratory or the field, every time he looks at something worthy of the record, he trips the shutter and in it goes, without even an audible click. . . .

I love Bush's description of the memex. The image he conjures is like something straight out of a Jules Verne novel. I envision a

luxurious mahogany desk festooned with brass push-buttons, le-
vers, and translucent screens. I can just hear the muffled clickity-
clicking of mechanical registers crunching numbers deep inside
the casing. But even though most of Bush's hardware suggestions
are now obsolete, the antiquated trappings belie the sheer bril-
liance of his prescience. Bush's desk with storage, screens, keyboard, stylus, and platen is the equivalent of today's desktop PC
with a microphone, multiple monitors, and a scanner. Add in a
tablet PC and you gain pen-based input. And sub-walnut-size cameras are now affordable and plentiful. Just about all new cell phones
and laptops come with one built in, and they can also be bought
and worn on their own.

Bush was writing with scientists in mind. "There is a growing
mountain of research," he lamented. "We are being bogged down
today as specialization extends. The investigator is staggered by
the findings and conclusions of thousands of other workers—
conclusions which he cannot find time to grasp, much less to
remember."

But he also realized that quantity was not the core problem.
"The difficulty," he wrote, "seems to be not so much that we publish unduly . . . but rather that publication has been extended far
beyond our present ability to make real use of the record . . . [It]
must be continuously extended, it must be stored, and above all it
must be consulted."

Bush wanted to free his fellow scientists from the drudgery of
searching and cross-referencing their books, journals, and notes so
that they could focus more on the creative side of their work.

"Creative thought and essentially repetitive thought are very
different things," said Bush. "For the latter there are, and may be,
powerful mechanical aids."

Bush did not want any scientist to worry about running out

of space in his or her storage unit, which would imply having to discard items that might later prove useful. Memex was to have infinite storage. "If the user inserted five thousand pages of material a day it would take him hundreds of years to fill the repository, so that he can be profligate and enter material freely," he wrote.

Memex would allow the scientist to annotate any item in the collection by speech or writing. Bush also wanted to support the way our minds work in associating one idea with another. He contrasted existing data storage with biological memory:

> When data of any sort are placed in storage, they are filed alphabetically or numerically, and information is found (when it is) by tracing it down from subclass to subclass. It can be in only one place, unless duplicates are used; one has to have rules as to which path will locate it, and the rules are cumbersome. Having found one item, moreover, one has to emerge from the system and reenter on a new path.
>
> The human mind does not work that way. It operates by association. With one item in its grasp, it snaps instantly to the next that is suggested by the association of thoughts, in accordance with some intricate web of trails carried by the cells of the brain. It has other characteristics, of course; trails that are not frequently followed are prone to fade, items are not fully permanent, memory is transitory. Yet the speed of action, the intricacy of trails, the detail of mental pictures, is awe-inspiring beyond all else in nature.

Bush hoped that "[selection] by association, rather than indexing, may yet be mechanized." To this end, he proposed that "trails" be created, connecting one document with the next in a sequence that could be followed again later. Trails could be given names, were something that you could share with your friends, and all the links were two-way. (The familiar hyperlinks on the World Wide Web are only partially realized trails. They are one-way, and are not grouped or named.)

Bush's memex was inspirational. The time was ripe to realize his dream—and to extend it far beyond the realm of scientific research into the lives of everyone.

A REAL MEMEX

We named our research project MyLifeBits, and adopted memex as its minimal requirement. Our goals were twofold:

1. To create software for lifelogging, and the subsequent recall and usage of one's e-memories. We wanted software to record a diverse array of information about one's life and activities, from a variety of sources and devices, and to do so as easily, as unobtrusively, and as automatically as possible. The software would have to give people powerful tools for searching, organizing, annotating, and pattern-mining their ultimately huge e-memories.
2. To identify the benefits, drawbacks, technical issues, sticking points, and usability of Total Recall in real life. We wanted to try it out (as much as we could) and see what it was like.

Since 2001 I have been serving as the primary test subject, but Jim Gemmell is also an avowed user, while Roger and Vicki have tried out numerous aspects of it in real life. A number of universities also have used our software and have experimented with it.

MyLifeBits is not a commercial product; it is a research project. In fact, MyLifeBits software is not a single application. It is a prototypical suite of applications, and a storage system that blends files with a database. You won't see Microsoft eventually ship MyLife-Bits version 1.0. Instead, you will gradually see more and more of the kinds of things done in MyLifeBits also being done in operating systems and in applications.

Our aim was to preview and to help lay the groundwork for the Total Recall systems that are coming soon—very soon—and that by 2020 will be as commonplace as Web browsers and cell phones are today. A few early steps in the Total Recall revolution have already hit the market. These include Evernote, reQall, OneNote, Google's Web history, and support for desktop search in operating systems. But as this book is being written these products remain small discrete solutions within a much larger puzzle.

HOW TO ORGANIZE AN E-MEMORY

Back in 2001 I could see we still had a lot of basic things to figure out about how to store and organize my data. We had just my sixteen gigabytes of documents and photographs loaded in my imperfect classification hierarchy of folders, and we had no good way to search them, sort them, annotate them, link them together into Bush's "trails," or analyze them for patterns and trends.

The files-and-folders method of organizing data is a fundamental feature of all modern operating systems such as Windows, Macintosh, and Linux. File-and-folder hierarchies, even when stored digitally, suffer from the same basic limitation as libraries once did: Each book can exist only in one place, filed under one category. But an item might properly belong to several categories, or hundreds. *A Brief History of Time* is a physics book, but it's also a book by Stephen Hawking, it was a best-seller, it talked about black holes, and it was published in 1988. You could easily come up with dozens of other attributes that would be perfectly legitimate criteria for tracking down *A Brief History of Time* and for sorting and grouping it with other books (and for that matter, for sorting and grouping it with other media of any kind: with lecture recordings, with songs, with articles, with pictures, with old news footage).

The MyLifeBits project ran into the problem like this: Logically, my Eagles folder should have been stored in both my Fun folder and my Animals folder, but in practice I had to make an arbitrary choice. And often, if I wanted to find some half-remembered piece of data again, I'd have to hunt for it in various folders, asking myself: *If I were me, where would I file it right now?*

Librarians have been familiar with this restriction for centuries. A copy of a book can only be on one shelf in just one section, often determined by the Dewey decimal system of book topics. So they created paper card catalogs, where a card was a surrogate for a book. Now the book—or at least its surrogate card—could be filed in more than one place. Dewey might have it placed in the physics section, but it could also be in the title card catalog, filed alphabetically by its title, as well as being filed alphabetically by author in the author card catalog. For your convenience, the Dewey subject index would be duplicated in the card catalog also, allowing you to flip through cards with your fingers rather than hiking down the aisles.

So here I was, with a system that was *worse* than a library with paper card catalogs. I was like a librarian who was not allowed to have a card catalog. Jim Gray, who is widely celebrated as a pioneer, even a founding father, of database design, shook his head over me.

"You need to use a database, Gordon. When are you going to listen to me?" he would ask.

I was resistant. "We don't need no stinkin' databases," I'd reply.

My resistance stemmed from my first experiences with databases back in the 1970s. Back then databases were still new and much hyped—they were also space-hogging and hard to use. I knew they'd been improved in the time since, but I'd heard enough horror stories over the years to keep my prejudices well nourished. Also, I wasn't clear on exactly what I wanted out of even a well-behaved one. But, it turned out Jim Gray was right—as usual.

A database is a program for storing and retrieving large collections of interrelated information. Modern databases let you very quickly retrieve all the records with a given attribute. You can rapidly sort, sift, and combine information in just about any way you can imagine. There was once a slight technical distinction to be made between how a database could index and look up records and full-text retrieval of documents, but by now databases have subsumed full-text search; they are happy to store documents and perform Google-like retrieval.

In his memex paper, Bush had expressed hope that the search algorithms of the future would be better than simple index-lookup on some attribute like author or date. He held up the human brain's associative memory as the ideal. In an associative network, items are linked together by contingency in time and space, by similarity, by context, and by usefulness. There are often numerous paths to each item.

Bush was right that trails and associative linking were critical components of an effective e-memory machine. But his dismissal of indexing was one of his rare failures of imagination. In his day, indexing meant alphabetical lookup in a predefined, noncomprehensive list of topics or keywords, as in a library card catalog. With the speed of modern computers, it has become possible to index every single word and phrase in every document and to search all of them in an instant. When indices are so comprehensive, and

lookup by the index instantaneous, then indexing is actually the
mechanism by which associative memory becomes possible.

The MyLifeBits research project revealed that any system that
aspires to be sold as an e-memory machine in the age of Total Recall will have to use a database storage engine, including full-text
indexing. Only a database will allow you to create two-way links
between items (including annotations) and to regroup and recat-
egorize items and collections of items in an open-ended fashion.
Only full-text indexing will give you keyword access to all of your
e-memory.

With MyLifeBits we could find all items that share a certain
property, such as having the same creation date, or having been
edited during a particular meeting, or having been viewed within
a certain span of hours after a particular phone call.

To make a database-style system work, we needed to include
what is called metadata, or "data about data." Metadata is essentially digital annotation about a file or other software object. Metadata may be embedded inside a file, or it may be "attached" to it
from the outside. Conceptually, it's a bit like a sticky label on a
manila folder that characterizes its contents.

Your computer's operating system keeps a little metadata on
each file for bookkeeping, such as its creation date, the date last
modified, the size of the file, and who is allowed to access it. Certain file types support additional metadata. For instance, the Microsoft Word document I am typing in right now lets me enter
author, title, subject, keywords, category, status, and comments.
Pictures in JPEG format can record things like the date taken,
location, camera model, focal length, f-stop, and exposure time.
Nearly all music formats include artist, album, composer, genre,
and length.

Some of this metadata gets filled in automatically. Digital cameras fill in the JPEG fields when they take a picture. CD-ripping programs look up the album information on the Web and populate the metadata for each song. In contrast, the metadata for Microsoft Office documents must be manually entered. Your name, which you were prompted for when you installed Office, is prepopulated as the author in new documents, but everything is blank—and tends to stay that way. Many of these manual-entry attributes will remain blank until we have systems more like MyLifeBits so that there is an actual payback for doing the work of entry.

One kind of metadata attribute that is getting a lot of attention these days is called a tag. A tag is simply a single word or short phrase. Tags aren't much different than the keywords attribute I have for this Word document. But they are creating a stir because there are some great photo applications, such as Flickr, that make tags easy to create and very useful for finding things again. You can add any number of tags to a file. For example, I have a photo of myself at age ten, with my four-year-old sister Sharon and my pony Snippy, who liked to bite. I kept that photo in a scrapbook for most of my life. It existed on one page only, and I was the only person who could find it quickly. But by putting the image into a database, I can tag it in any way I might find useful: Gordon, Sharon, Snippy, 1944, black-and-white, Missouri. Thereafter, I can use the tags for searching and sorting.

We also needed to be able to forge links—Bush's "trails"— between any items or collections of items in my database. For instance, I wanted to be able to link some photographs to an entry in my calendar, to indicate that they were photos of that event. Or, if I record some audio of me talking about a photo, I want to be able to link the audio to the photo, so it is clear that it is a comment about the photo.

So Jim Gray and some other colleagues convinced us to take

the plunge, and we created a database to hold all our files and other information. It was great. We could still view my data using my original folder-based organization scheme if we wanted, but that became just one of myriad ways we could view it. We could group items into what we called "collections," and each item could belong to more than one collection.

We worked hard on annotations and metadata, making them automatic whenever possible, and otherwise trying to make it quick and easy at any moment to add information. For instance, I could select a bunch of items and then type a comment about them at any time. If I didn't feel like typing, I could click a button and just say the comment. Silence would automatically be stripped out of the beginning and end of the comment so I could be relaxed about hitting the start and stop buttons. I could also add ratings to any item. I could comment and rate Web pages from inside the browser. I could rate and comment on anything that came up on my screensaver.

Here's an example of what we can do with MyLifeBits, courtesy of a database design with good metadata and complete indexing. Say I'm trying to remember the name of a biotech entrepreneur I read about several years ago. I can't remember his name or his company or anything else specific enough for a standard search. What I do remember is that I read about him on the Web, the article involved biotech, it was between two and four years ago, I was at the office, and it was during a fairly long phone call with Jim Gemmell—say, ten minutes or longer. Those are pretty vague parameters, but they're enough for MyLifeBits to winnow the selection down to just a handful of archived Web pages. I quickly find the name I need.

In my old files-and-folder system, I had metadata, such as date of publication, name of person, et cetera, embedded in long file names. But in the database, we could have an actual publication-date field that I could use for sorting and searching.

Unfortunately, we didn't have the manpower to make all the existing programs out there work smoothly with our database, so we ended up having our database keep an eye on a regular file-and-folder system and stay synchronized with it. This gave the folder system more prominence than we would have liked, and made the overall system more fragile, but that was the reality of creating prototypes with limited resources. Thankfully, it was just a nuisance, not a serious roadblock. Our software still looked pretty much the same, and we still could learn from real-life experience with this kind of storage, no matter what was under the hood. We were up and running.

CAPTURE EVERYTHING, DISCARD NOTHING

While we were thinking through the memory organization problems, I continued capturing and saving more and more of my life bits. The project mantra had become: Capture everything, discard nothing.

We made it a goal to make capture as automatic as possible; otherwise I knew I just wouldn't capture enough. We enhanced my Web browser to record a copy of every Web page I visited—not just the URL that points to it, but a copy of everything on the page. The advantage of this is that it solves the problem of "link rot," the process in which hyperlinks gradually become invalid, one by one. Link rot happens for several reasons. Web sites sometimes restructure their content, change hosts, get bought up, or go extinct. Other Web sites make content free and viewable when it's fresh but disable the links after a few weeks. Another problem arises with sites that are continuously edited, such as political position papers and Wikipedia entries. Creating a copy of every page I visit in the exact form it had at the time circumvents all these problems. Furthermore, it is often easier to find a page from my collection of seen pages, rather than search for it out of the entire Web.

This page-logging can be turned off so that I can visit sites without having them go in my e-memory. However, with all the storage at my disposal, there's not much point. I literally can't surf the Web fast enough to incur a significant storage cost.

I also started recording all my instant messaging and saving all my e-mails, minus the spam (just like my paper, I want to keep what I actually read, not what marketers force into my in-box).

We set up hardware to record telephone calls in my office. If you call me, you will first hear a voice say, "Recording." This notifies you that the call is being recorded, as is required by California law (not to mention common courtesy). I can settle any dispute about what was said on a conference call by instantly retrieving the audio file. My alibi in court, if I ever need one, will be ironclad to the extent I can prove that I didn't fabricate it.

We started tracking all kinds of things: the number of mouse and keyboard clicks, every time a document was opened, every window shown on my PC screen, and the history of my music playback. We logged every search. I bought a GPS and started loading my location history into MyLifeBits.

We even experimented with recording radio and television shows. Digital video recorders (or DVRs) such as TiVo were just coming out, and we wondered what it would be like to keep everything when it came to TV. We built our own DVR and set it up with nearly two terabytes of storage—more than twenty times the capacity of the early DVRs. If you think your TV program guide is big, try wading through more than a thousand shows, all of which are actually interesting to you. And radio was a totally different experience. We recorded lots of National Public Radio shows, including *Prairie Home Companion, Car Talk,* and news. We played the audio back on a Pocket PC, so it was like a cross between TiVo and podcasting. Jim Gemmell learned that he fast-forwarded through all but fifteen minutes of a typical news hour.

But I quickly lost interest in TV and radio because such shows would soon be archived and available on demand. Having your own copy is not so special if you can just have it streamed to you through the ether anytime you please. It's still worthwhile to have your lifelog make a record of what you watched and when, but not to copy the program itself.

By October 2003, I still wasn't wearing the walnut-sized camera strapped to my forehead that Bush had predicted. But Lyndsay Williams, a colleague from the Microsoft Research Laboratory in Cambridge, England, had come up with something even more interesting. She called it the SenseCam. About the size of a cigarette pack that hangs from a cord around your neck, the SenseCam is a fisheye camera that takes pictures automatically. When it detects a change in light level it presumes you've passed through a door or otherwise changed your setting, and snaps a picture. When its passive infrared sensor detects the appearance of a warm body, it snaps a picture of whoever just came into view. An accelerometer lets the SenseCam know when to delay taking a picture to avoid motion blur. And of course, you can point the SenseCam and take photos at will rather than waiting for it to take the initiative.

Lyndsay once confided that one reason she developed the Sense-Cam was to find her misplaced eyeglasses. By scanning SenseCam images, she can find the last place she put them down.

One of my favorite examples of how the SenseCam enhances life comes from Cathal Gurrin, a lecturer at Dublin City University in Ireland. Cathal set out to perform a year-long experiment, wearing the SenseCam during all his waking hours. When the year was over, many people expected him to be glad to stop. In fact, he wouldn't give the SenseCam back. Cathal began wearing the SenseCam daily in June 2006 and, as I write, has worn the Sense-Cam for almost three years, acquiring over three million photos. Gurrin has a collection of his favorite photos rotating on a digital photo album on his desk which he shows off with the enthusiasm of a new parent with baby pictures. "Look," he says, "here's a picture of the first moment I met my girlfriend—not that I knew she'd become my girlfriend at the time."

A fun thing to do is to play back all the SenseCam images from a day or a week in rapid succession, which takes just a few minutes. Talk about your life flashing before your eyes! It's an amazing feeling to see your life on fast-forward like that.

I enjoyed taking the SenseCam on walkabouts with my GPS. I could later reconstruct my travels on an animated map, with pictures taken along the way to tell the story. The best series I did was an eight-hour trip along the Great Ocean Road in Australia and through a treetop walk in a rain forest.

My SenseCam has captured many special moments, especially at parties, lunches, and conference exhibits. I have a sequence of when I was admitted to the hospital for heart bypass surgery in July 2007. My partner, Sheridan, wore the camera as I was wheeled into the operating room.

THE DAY OF CARPE

By 2004, we were so excited about where MyLifeBits was taking us, and saw so much potential, that we wanted to encourage others to get involved. Jim Gemmell launched a workshop at ACM Multimedia 2004, a professional conference for computer scientists. The theme of the workshop, which was held annually for three years, was CARPE: Continuous Archival and Retrieval of Personal Experiences.

In 2005 we invited universities to submit proposals for research projects. We received eighty submissions and selected fourteen of them to receive money, SenseCams, and our software. I was thrilled at the wonderful results from academia, touching into many areas and ideas we would never have thought of, from helping disabled students to logging therapy sessions for stroke victims.

As of this writing I have 261 gigabytes of information saved on my main computer and about 100 gigabytes accessible in my cloud. I add about one gigabyte a month. This doesn't include continuous audio and video, but that's on the horizon.

The MyLifeBits software is far from perfect. The hardware right now is clunky enough that I don't use it all the time (I hate dealing with heaps of batteries and chargers!). But between MyLifeBits and the work of our colleagues in the research community, we believe we have a proof of concept. We've built and experienced enough to confidently endorse Total Recall.

We will be taking a tour of how Total Recall has affected my life so far and how it will affect your life, in ways direct and indirect, large and small, as e-memory becomes standard furniture in our daily lives. Before we get to the effects Total Recall can have on work, health, learning, and our personal relationships, we need to take a deeper look at what science can tell us about the meeting of e-memory and bio-memory, that stuff that resides in our heads.

CHAPTER 3

THE MEETING OF E-MEMORY AND BIO-MEMORY

I was invited to a birthday party for computer graphics pioneer Ivan Sutherland. Would I be willing to say a few words?

Great, I thought. *I can get up and tell them I run into Ivan about every five years, and we enjoy our conversations.*

Honestly, that was all that came to me. Then I entered Ivan's name into the search window in MyLifeBits, and to my surprise and relief, I immediately recalled emotionally evocative and intellectually intriguing details I had completely forgotten. Around 1963, when Ivan had been a first lieutenant stationed at the National Security Administration, he and I designed a display computer. A few years later, he had been instrumental in my becoming a professor at Carnegie Mellon University. Much more recently, Ivan had started opening his talks with "Gordon Bell and I have a friendly debate over whether I've wasted my life working on asynchronous logic." (Come to think of it, I hope he'll take my thousand-dollar bet that asynchronous logic will be widely recognized as a waste of time by 2020).

My biological memory had reduced my relationship with Ivan down to the humdrum, but my e-memory stepped in to restore the significance of our history, making it possible for me to compose a fitting toast for his birthday.

We all want better recall. The market for memory enhancement books, elixirs, computer programs, devices, and games is gigantic.

As people get older, they start to get paranoid about small memory lapses. When a forty-year-old misplaces his car keys, he feels annoyed. When a sixty-five-year-old loses the keys, he starts Googling about Alzheimer's disease. In his search he might read about another condition known as mild cognitive impairment, which afflicts as much as 5 percent of the population past the age of seventy. It's very real, and very scary.

The fear of oblivion before death is big enough to drive a $4.2 billion industry in medicinal herbs and supplements for memory enhancement. Health-food emporium shelves are stocked with herbs, micronutrients, antioxidants, tonics, supplements, and potions to boost your brainpower. Labels on bottles of coenzyme Q10, ginseng, ginkgo biloba, rosemary, and salvia promise to keep your mind nimble.

In 2007, the U.S. market for brain-fitness programs and "neurosoftware" was $225 million. Nintendo sells a product called Brain Age that claims it can help you "[get] the most out of your prefrontal cortex!" The software program MindFit combines cognitive assessment of more than a dozen different skills with a personalized training regimen based on that assessment. Dr. Michael Merzenich, a neuroscientist at the University of California in San Francisco, developed a set of computer tools called the Brain Fitness Program intended to increase processing speeds in aging brains. And for about ten dollars a month, you can subscribe to Web sites like Lumosity.com and Happy-Neuron.com to tap into a variety of cognitive training exercises.

You can also buy books on how to exercise your brain with games, puzzles, and memory tricks. You can learn about the biology and behavior of memory from books such as *In Search of Memory: The Emergence of a New Science of Mind* by Nobel Laureate Eric Kandel and *The Seven Sins of Memory: How the Mind Forgets and Remembers* by Harvard University psychologist Daniel Schacter.

For all I know, ginkgo biloba and the Brain Fitness Program will indeed improve your bio-memory. But the world of Total Recall promises something broader: a revolution.

THE INESCAPABLE FRAILTY OF BIO-MEMORY

The memories in our brains are stored as patterns of connections between neurons, or nerve cells. Computers store information in a series of microscopic switches turned on or off. Brains and computers both store information, manipulate it, and use it to decide between courses of action. For these reasons we say that both systems have "memory," but this similarity only holds up in the first approximation. Scratch the surface and you find vast differences between biological and digital memory.

To the owner of a human brain, memory feels like a single resource. It turns out this feeling is an illusion. Scientists who study biological memory describe three distinct systems:

- Procedural memory, sometimes known as muscle memory, is for physical skills such as riding a bike, ballet dancing, typing.
- Semantic memory encodes meanings, definitions, and concepts—facts that you know that aren't rooted in time or place, such as "A cat has four legs" or "The capital of Japan is Tokyo."

- Episodic memory, sometimes known as autobiographical memory, encodes experiences from your past. This is what allows you to know about and reexperience the things that have happened in your life, such as the time you sprained your ankle at the playground and your father bought you an ice-cream sundae to make you feel better, or the shower you took an hour ago.

Nothing is coming soon that can help us with our procedural memory. But our semantic and episodic bio-memories can and will be extended by our e-memories.

We all know that biological memory is fallible, but it's still extremely unnerving to learn just how true this is. As neuroscientists have shown, episodic memories feel a lot more fleshed out and precise than they really are. Unlike computers, brains aren't all that great at faithfully storing masses of detail. What brains are best at storing are patterns, meanings, and gestalts. The act of remembering an event from your past is less like playing back a mental videotape in your mind's home theater system than it is like telling a story based on a few relevant facts.

In an age of Total Recall, anything, even everything, is easily recorded accurately into your e-memory. Your brain can't do this. When it lays down a new memory of an experience, what it actually encodes is a sparse constellation of authentic details and salient junctures. When your brain retrieves the memory later, it uses that constellation as a scaffold for reconstructing the original experience. As the memory plays out in your mind you may have the strong impression that it's a high-fidelity record, but only a few of its contents are truly accurate. The rest of it is like a bunch of props, backdrops, casting extras, and stock footage.

When a friend tells you a five-minute-long humorous story, the memory you come away with isn't the exact sequence of words he uttered. When you repeat the "same" story to your friends at work on Monday, what you actually do is reconstruct it in your own way according to the same pattern and meaning. You follow the overall map provided by those key junctures you memorized, but you freely embellish and fill in any gaps to make the story flow smoothly between them. You might repeat verbatim a few key bits of the original telling, but most of the word choices are yours. Generally, all I can remember of a joke is the great punch line that made me laugh, and I have to reinvent the rest in order to share it.

And it gets even stranger. Sometimes a feature that was confabulated during one act of remembering gets reremembered during the next act. In the process, the confabulation can become a permanent element of the bio-memory. Here's how the eminent neuroscientist Joseph LeDoux sees it:

> Like many scientists in the field of memory, I used to think that a memory is something stored in the brain and then accessed when used. Then, in 2000, a researcher in my lab [convinced us] that our usual way of thinking was wrong. In a nutshell . . . each time a memory is used, it has to be restored as a new memory in order to be accessible later. The old memory is either not there or is inaccessible. In short, your memory about something is only as good as your last memory about it. This is why people who witness crimes testify about what they read in the paper rather than what they witnessed.

False memories can have tragic consequences. In the late 1980s and early 1990s, thousands of families were ripped apart when adult children claimed that they had recovered long-repressed memories of sexual molestation when they were little. It turned out many of these "memories" had been coaxed into being by gullible, credulous therapists who hadn't realized what they were doing.

Most of our memories are not grossly altered as our brain repeatedly remembers them, but all of us harbor at least some memories that have been radically revised, and all of our memories are susceptible to gradual mutation and drift.

That is about to change.

E-MEMORY TRANSFORMS BIO-MEMORY

Biological memory is subjective, patchy, emotion-tinged, ego-filtered, impressionistic, and mutable. Digital memory is objective, dispassionate, prosaic, and unforgivingly accurate.

In our brains, memory, attention, and emotion conspire to warp, compress, and edit time and life experience in many ways. A video camera, the eye of an e-memory, in contrast, never blinks or winces, never drifts into a daydream or does a double take. A camera will record an hour of pedestrian crosswalk traffic with the same fidelity as it will witness an hour of bloody genocide.

E-memory will be the fact checker for those meanings, definitions, and concepts in our semantic memories. You probably already use Google or Wikipedia to look things like this up, when you can. But not everything you know is easy to find on the Web, or may not even be there. It will be there in your e-memory. And it will be easier to find, because you will be searching just your own memories, not the whole Web. I often find it is faster to use MyLifeBits to track down obscure facts I know I've been exposed to before but can't recall directly, simply because I'll often remember when or where or from whom I heard the thing I'm trying to recall.

Everyone knows the anxiety and frustration of not being able to remember someone's name. With MyLifeBits I often track down a name using clues I do remember. I recently wanted to find the name of a fellow who nearly contributed to the Computer History Museum in 1983. I recalled the company he worked for. I thought he came to a lecture at the museum that same year. I wondered if his name was on the list of attendees, a copy of which I had kept in a box for years and then scanned. . . . Yes! In general, if I know that my e-memory has it, I will usually find it within a minute or two.

E-memory will become vital to our episodic memory. As you live your life, your personal devices will capture whatever you decide to record. Bio-memories fade, vanish, merge, and mutate with time, but your digital memories are unchanging. And e-memories will contain an unprecedented level of detail. With my bio-memory, I struggle to recall exactly when I was in San Francisco last year. With my GPS logged, I can recall the exact time of my walking down each individual street in San Francisco.

Total Recall will change how we think about our lives. It will also change how we feel about our lives.

Consider just one photo that popped up on my screensaver while I worked on this book. I glanced at it and was transported back to my fourth birthday in 1938. Mother told me I could invite anyone I wanted to my party, and so I did. We were an eclectic bunch, all eighteen of us ranging in age from two to fourteen. I'm in the middle front with a large sheet cake on my lap. It's obvious that I've got more important things on my mind than getting my picture taken, like sticking my pudgy fingers into the creamy frosting, which although white, hid a middle that was pure devil's food.

The faces in the picture spark recollections, like the really cute three-year-old girl from across the street, sitting in the front row with me. When my sister was born two years later, I picked the name Sharon, after that little girl, my first sweetheart, Sharon Lee.

We were framed by my older teenage cousins, one with his hands in his pockets, looking ever so cool, while the other was praying for the photograph to be over. Glancing at each face, I was struck by one in particular. His name was Joe Bill, the minister's son. He died a year later at the age of ten and it bothers me as much today as it did then.

This single picture from 1938 initiates an avalanche of memories, each connected to the other, via associations established in my brain decades ago. They elicit feelings of pleasure and sadness. Each strand tugs on a dozen others, all of them connected into the vast web of memories that make me uniquely me.

I recall a research demonstration at Microsoft headquarters of a huge grid of LCD monitors, three high and six wide, all filled with a time line displaying photos from my life. I stood transfixed for several minutes in front of this biographical vista, soaking in the perspectives and the details. Seeing so much of my life, all at once, was profoundly moving.

E-memories will not be trapped back in cabinets and shoe boxes. They will be on our end tables and walls. They will follow us on our travels. They will keep us company, showing us friendly faces, letting us hear cherished voices. E-memory will be an intimate extension of bio-memory. And change it into something new.

RECALLING WHAT MATTERS

I don't know about you, but sometimes I'm absentminded. I forget where I put things. Sometimes, coming out of the airport terminal, I have a momentary flash of panic. Where did I park the car? Was it level one or two? I hate getting home from the grocery store, looking up at that burned-out lightbulb, and realizing that I forgot—again!—to buy a replacement.

Once I left my notebook computer containing most of my e-memory on the security table at San Francisco International Airport. I dashed back, my heart racing dangerously, wondering if someone had walked off with a digital copy of my life. Thankfully, it was still there. Then I forgot the computer again at the Dulles Airport security, and didn't realize my mistake until I had boarded the plane and it was too late to go back. I managed to have it overnighted to me for $150, and all I could think was that I would gladly have paid many times that amount to ensure no one else had my data. More than a half million of my fellow Americans also left their computers at checkpoints in 2008.

A busy person may be plagued by absentmindedness, simply because he has a lot on his mind. You forgot to bring home the milk, because by the time you got to the grocery store you were thinking about the items needed for your pets. You forgot your lunch meeting, because you had just gotten off the phone with a colleague and were engrossed with new ideas for your next project.

Reminders must be made at the appropriate time; it is no good having a shopping list that is back at your office while you are in the grocery store. A reminder to make a phone call while you are wasting time in rush-hour traffic would be priceless. Likewise, you must be able to create reminders anytime, anyplace, or they may be lost. If you think of something that needs doing while driving, it will not suffice to have to wait a half hour to get home and write it down. Probably by then you will have thought of three other things that need doing and forgotten at least one.

This is why e-memory will be in the cloud, accessible anywhere, anytime. We want to be able to type or speak notes and to-do items whenever they occur to us. The to-do items will be associated with a time, place, or mode of activity. For example, the task "Buy milk" is associated with the grocery store (a place). The task "E-mail Catherine" is associated with using your computer (an activity). The task "Pick up Suzy at 4:00" is associated with a time. If you are struck by the thought that a cell phone is capable of knowing time and place, and can input text, voice, and pictures, then you realize how close we are to the reality of this vision. All we need is a little more software that can understand such things as milk being available at grocery stores.

In addition to giving you all the right reminders, it will not be too long before your e-memories will fill in your other absent-minded gaps. Your increasingly location-aware cell phone will remind you where you parked your car. You will track where you have left things like your glasses, either by noting where your devices last detected their RFID tag, or by taking pictures of them. When your mind is absent, your e-memory will always be there.

Having too much on my mind doesn't just make me absent-minded; it can make me feel mentally cluttered, impeding my productivity. David Allen's popular book and seminar series *Getting Things Done* stands on the central premise that we are hindered by mental clutter:

> First of all, if it's on your mind, your mind isn't clear. Anything you consider unfinished in any way must be captured in a trusted system outside your mind. . . .

Unlike your bio-memory, your e-memory will never be overwhelmed. Total Recall software can make sure you are protected from clutter. For instance, if I were to show you all my 150,000 recorded Web pages, you would see that nearly half are duplicates, or are near duplicates that differ by just a tiny amount from some other page. A natural reaction would be to delete all the near duplicates to eliminate clutter. However, e-memory never needs to suffer from clutter; only a poor recall interface looks cluttered. Good recall software could simply group all the near duplicates together and show a single representative in response to your queries. Suppose I have repeatedly gone to the same Web page and the only thing that changed on it was the advertisements. Most of the time, I just want to see one page, hiding away the clutter of all the extra copies with inconsequential differences. But on that one day that I want to recall an ad for 50 percent off a new GPS, suddenly the differences are very consequential—and I've still got them.

Total Recall software will hide away all the clutter as if it had been discarded, but whenever you actually want it, it will be at your fingertips. You will have all the advantages of complete retention, with none of the downside of clutter.

And e-memories are not depressed by dealing mostly in the mundane. One thing you'll notice when perusing a lifelog is the sheer banality of 99 percent of life. Television producers periodically demonstrate this when they have a camera crew live with a family for a long time or Reality TV records a group surviving on an island twenty-four/seven. You will very quickly come to appreciate just how mind-numbingly dull, trite, predictable, tedious, and prosaic most of our life moments really are. Life as it appears in objective playback is tedium to the dullth power. But that's not a problem for an e-memory. You know that you'll never want or need to look back at virtually all of it ever again. You'll also know that nothing important will be missed—just as Cathal Gurrin has that special moment of first meeting his girlfriend.

The team that Cathal works with at Dublin City University has attacked the banality of lifelogging by creating software that looks for novelty. It works something like this: Suppose your GPS says you were at the same place as usual this morning for breakfast, and the images look very similar to those taken most other mornings.

Then it was probably the mundane, same old breakfast at home as usual. On the other hand, if you were at an unusual place for breakfast, or more faces were around the table than usual, that is more interesting. Cathal and his colleagues take thousands of SenseCam images, and boil them down to a presentation that highlights the unusual. The mundane is still there, but tucked away so that it doesn't clutter up what is interesting.

Total Recall software will get better and better at spotting the interesting moments for you as we acquire more data such as your pulse, the pitch and volume of your voice, or even the brain waves you are emitting. Researchers at the University of Tokyo have already demonstrated a baseball cap that records both video and alpha waves from your brain. Based on the alpha waves, they can do a pretty good job of guessing what segments of the video were interesting.

For an e-memory, there's no drawback to capturing the long stretches of banality that comprise most of life, there are only potential benefits. Go ahead and fill it with et ceteras, so ons, and ad nauseams. The more you record, the better.

Your mind can be freed from mundane memorization. Let your e-memory remember each detail, and show you the average, the maximum, the chart, the patterns, or the unusual. Then when you decide that one particular area bears further investigation, you can recall all the gory details—perhaps the actual data points, or some additional photos. Knowing that your e-memory has the task of perfectly remembering allows you to concentrate on more interesting things.

E-memories not only relieve you of memorizing what you don't care much about, they can also help your bio-memory remember what counts.

My friend Sunil Vemuri is the CTO of reQall, a really fascinating memory-aid product. I use reQall to create reminders and notes for myself. I call up their phone number and say, "Add." There's a beep and then I make some comment, perhaps, "Book the flights for my vacation." I can retrieve this reminder by calling in later and saying, "Recall," instead of "Add." I also get an e-mail, with a recording of what I said, and even a transcription of what I said. Besides using the phone and e-mail, I can use other interfaces like a Web browser and instant messenger. Being able to create and retrieve memos in all these ways is very powerful. I even use the phone to create dear-diary entries: I can be driving along, place a call on my cell phone, and tell a little story that ends up transcribed in my e-mail and also with my own voice recorded.

All that was exciting enough to get me to join reQall's board, but Sunil expanded my vision of what is possible when he visited Jim Gemmell a little while ago. They sat down in Gemmell's office and were chatting about all kinds of Total Recall ideas (when Sunil was still a graduate student at MIT he got involved in the CARPE research community, so he's been a coconspirator with us for years). In the course of the conversation, Sunil made reference to Gemmell's family, and even what school Gemmell's oldest son attends. Gemmell had only mentioned this once, and that was more than a year ago, so he was astonished at Sunil's memory. But Sunil gave up his secret later.

"Do you know how I remembered about your family and your son's school?" he asked.

Gemmell shook his head and Sunil went on. "After our last meeting, I called reQall and spoke some notes about our meeting, including those facts about your family. Now, I also have reQall programmed to play random facts to me every so often. Since our last meeting, I've heard those facts about your family a couple of times, so now I remember them."

This kind of memory refresh is the driving factor behind an-

other product called SuperMemo. Instead of just randomly reminding you, SuperMemo considers the typical pattern of memory loss.
Cognitive scientists have measured how memories typically fade,
and can plot the odds of your forgetting something after first hearing it, after one reminder, after two reminders, and so on. Supermemo intervenes when the odds of your losing the memory reach
a certain level, say 15 percent. For instance, two days after hearing
something, you might be projected to have a 15 percent chance of
forgetting the fact, so you are reminded. Eight days later, you again
are projected to have more than 15 percent odds of forgetting, so
you are reminded again, and so on, with the time between reminders growing longer and longer.

RETAINING PAINFUL MEMORIES

When I give talks about MyLifeBits, someone in the audience usually says something like, "But isn't forgetting a good thing sometimes? Isn't the idea of recording our lives in excruciating detail
actually a rather bad idea? Don't we need to forget?"

Everyone has experienced embarrassing moments he'd rather
forget. You're the quarterback who ran the wrong way down the
football field. You called your lover by the name of an ex-flame.
One of my most embarrassing memories was shown on the Business Channel in 1983. I was one of a dozen company founders sitting at a table, explaining to the press our plan of merging with
another company to turn our fortunes around. There was only
one hitch with this ill-fated plan: We had nothing to sell. I have
nothing but scorn for product announcements without an actual
product, and here I was in the thick of it. Every time I think of it
I feel an echo of the original horror I felt clamoring around in my
gut—which is why I try not to think about it very much.

But you can easily avoid replaying such memories. And who knows? Maybe someday when you're old and gray and retired and you look back across your life with the expanded perspective that often comes in the winter of life, you might actually be able to look back at your old fumbles, gaffes, faux pas, and humiliations and gain closure on them, forgive yourself for them, even laugh at the ultimately petty little anxieties that used to seem so serious. In a perverse way I would love to have a copy of the Business Channel tape just to see whether I was crawling under the table just as I wish to remember the moment. I know I don't mind watching a 1972 video of me making a shortsighted prediction about where computing was headed.

But what about the really, truly bad memories? Not the harmless embarrassments that still make you blush, but the ones that are so compromising or potentially harmful—to your reputation, to your loved ones, or to your own sanity—that you just can't abide the thought of keeping them?

What about a woman who's been abused by her husband? When she finally escapes him and gets help putting a new life together, what could be more unwelcome than digital records of the horror that had been her life? The last thing she wants to do is relive the insults, the threats, the cat-and-mouse mind games, the screams, the beatings, and the bruises. It would only be natural for her to want to delete every last bit of it.

What about a young man who makes some bad decisions in high school? He makes the wrong friends, starts experimenting with drugs, and ends up in a stoned stupor in the backseat of a stolen car. He is arrested, and the experience with the juvenile justice system scares him straight. The state expunges his juvenile record and he goes on to become a law-abiding citizen with a family and a profession. In a world with no records, he could easily leave the past behind. In a world where most things are recorded and saved, would he have the same chance?

Nevertheless, I still advocate keeping everything, even the worst of it. They are your e-memories; you control the keys to them. Rather than erase them, you can seal them up. You can put a lock on those events you'd like to forget and never open them up again. What you really want to prevent in these cases is unwanted *recall*, not retention.

Imagine the abused woman has audio and video recordings of the abuse she endured. She has escaped, gotten treatment, and is living in a new city without fear. Her recordings can easily be locked so that they'll never come up in regular interactions with her e-memory. But she may want keep them for legal proceedings. Or she may want to share them with future therapists.

Imagine the young man who was arrested, now grown older and involved in a community effort to block a new commercial development. His opponents start to circulate stories that he was a hard-core criminal in his youth, with gang connections lasting to this day. He finds it in his interest to show his youth record to defend himself against this slander.

Our impulse to hit the delete key may not be the right move to lock away the past. Daniel Schacter advises that "confronting, disclosing, and integrating those experiences we would most like to forget is the most effective counter to [unwanted recall]."

LOST BUT NOT FORGOTTEN

I have personal experience with unwanted recall using MyLife-Bits. On Sunday January 28, 2007, my manager and dear friend Jim Gray took his forty-foot yacht, *Tenacious,* on a solo sailing trip out to the Farallon Islands near San Francisco. He went to scatter his mother's ashes in the wild seascape around the rugged islands.

But Jim never returned. Despite clear weather and no signs of distress from his well-equipped yacht, Jim mysteriously vanished. A massive three-week search did not produce a single clue as to what had happened. As *The New York Times* reported, "A veritable Who's Who of computer scientists from Google, Amazon, Microsoft, NASA, and universities across the country spent sleepless nights writing ad hoc software, creating a blog, and reconfiguring satellite images so that dozens of volunteers could pore over them, searching for a speck of red hull and white deck among a sea of gray pixels."

For several months after Jim's disappearance, I was deeply disturbed every time a picture of him came up on my screensaver. I avoided going into the office where we had worked. I was overcome with emotion. It was too painful for me to see his smiling face. Some people in my situation would have deleted those pictures, hoping it would bring relief or catharsis. But with time, my frame of mind changed. The same pictures now bring back happy memories and nourish my spirit. I'm glad I still have so many images of my old friend.

I used some of the pictures when I was asked to speak at Jim's memorial service. I knew my emotions would get the better of me on the day, so I planned to get myself out of an actual speaking role. I used Microsoft's Movie Maker to load my collected snapshots of Jim. The software allowed me to drag and drop in special effects, such as fading from one photograph to the next. I wrote a script that I voiced-over the pictures, and concluded with Pink Floyd's "Wish You Were Here."

It took more than a week to create. When it was shown at the service, tears welled up in eyes all over the auditorium, including mine. My e-memory, working alongside my bio-memory, told the story of Jim and me, how our paths had crossed, and how rich my life became because of him.

There are many ways to create stories. My mother spent several months typing the story of her family, the Gordons (I'm named after her). It's about twenty-five pages long and it's chock-ful of stories about her family and about growing up as wonderful twentieth-century inventions like the automobile, electricity, and telephony became a part of her life—things we would have never known without those twenty-five pages. There's only one thing wrong with it. I should have had her record it in her own voice.

That's the kind of omission that will soon be a thing of the past.

PART TWO

PART TWO

CHAPTER 4

WORK

I like working. About six thousand people worked for me when I was head of research and development at Digital Equipment Corporation (DEC) from 1972 to 1983, before it became a part of Hewlett-Packard. I've been involved in more than a hundred start-up companies—that's an average of about four a year since I started. I've served on government panels, offered my thoughts in think tanks, given talks to all kinds of audiences, and met with countless young entrepreneurs to hear pitches for my involvement in their great new ventures.

As an angel investor, my interests are both literally and figuratively all over the map, so it is a juggling act keeping up on all the different technologies, business plans, and people whizzing in and out of my orbit. I've been vowing for a couple of decades to reduce my travel, but still end up with more than fifty thousand air miles every year (at five hundred miles per hour that is two hours per week in the air—far from extreme in our frequent-flyer age). And then there's my day job, working to advance Microsoft's technology for e-memories, Total Recall, and data-intensive science. So my calendar is usually full, and it's a challenge to get everything done.

At DEC, I used to get out of the office and go home when I wanted to actually accomplish anything because I was so inundated with interaction at headquarters. My schedule was kept in calendar notebooks, in pencil. My home and work offices used to be crammed with filing cabinets, bookshelves, and great teetering heaps of paper covering most of the available horizontal space. It wasn't actually as chaotic as it probably looked to others. I had a system in place—I am an engineer, after all—but still, the thought of ever going back to that way of organizing things sends a different kind of shiver down my spine.

In those days I was a "piler"—I created a pile of paper for each problem or topic that came my way. I wish I had a photo of the pile wall of my home office in Lincoln, Massachusetts, where I lived. The wall was roughly twenty-five feet long. It had six rows of shelves, providing room for around two hundred problem piles. When a memo, report, article, or something relevant to the issue came in, I merely stacked it on the right pile. Items were filed archaeologically, that is, deposited in layers over time. They were retrieved by lifting stacks of paper to reveal the archaeological era of the paper I wanted. I've met a lot of fellow pilers over the years, and seen some impressively high piles, especially in universities.

Total Recall will make it possible to deal with a prolific and even hectic work pace, far above our current expectations—and still remain sane. It will help make you more productive, whether you are a busy traveling salesperson or a parent frantically chauffeuring your kids between school and activities.

Being essentially paperless will be a big factor in this improvement. Instead of archaeological digging, a few keystrokes and mouse clicks will get what you need. Paperless offices are far more pleasant, and somehow calming.

Total Recall will also give everyone an incredible sense of freedom. Travel anywhere, anytime, and maintain complete access to every detail concerning your enterprise. I've experienced a taste of these benefits already, but work lives in the coming generation will become amazingly more powerful generators of prosperity and satisfaction.

THE NEW JOB

When you start at a new job, it can take a while to get up to speed and learn the ropes. How intensive and how crucial this process is varies widely. It might not be too big a deal in the case of, say, a new waitress at a theme café, who may take a few weeks to figure out what the style of the place calls for and the predictable orders of various regular patrons. At the other extreme it may have enormous importance, such as when a new president takes the helm of the world's mightiest nation.

I don't recommend this movie for students of political science, but in *National Treasure 2* there is a presidential "book of secrets." The book supposedly contains secrets for a president's eyes only, and is passed on to each new occupant of the White House. While no such book actually exists, at least as far as I know, there is clearly a vast body of knowledge that must pass from one POTUS to the next. The news media loves to run stories on the recently elected politician who doesn't know X concerning his or her new duties. Just think of the enormous body of knowledge that an aspiring president is expected to have on the tip of the tongue.

In between the president and the waitress are a million other jobs that have memories to pass on. Total Recall will break new

ground in the effectiveness of transferring memories from one occupant of a position to the next.

In order to tolerate staff turnover, many large organizations
are structured around clearly defined functions and operating
procedures for each member of the team. There is no better example than the military, whose constant movement of personnel
demands an extremely modular approach at all levels of its com-
mand structure. Soldiers and officers routinely rotate in and out
of positions on their tours of duty, and even homeland bases and
training facilities regularly shuffle their staff. Whether you are a
new base commander, quartermaster, or front-line soldier, you are
expected to drop into a position and be effective the moment your
boots hit the ground. National defense, with such well-defined
roles, is a fertile area for the application of Total Recall.

In early 2003, program director Doug Gage and some of his
colleagues from the Defense Advanced Research Projects Agency
(DARPA) came out to San Francisco to meet with Jim Gemmell
and me. They were interested in MyLifeBits as a model for a research program they were hatching called LifeLog. They had already held a LifeLog workshop and decided they were interested
in a system that "captures, stores, and makes accessible the flow
of one person's experience in and interactions with the world" and
that "can be applied to a wide spectrum of associate/assistant systems to allow the system to 'understand' the user's state based on
knowledge of the user's history (timeline, routines, habits, etc.), in
order to make the user more effective in a wide variety of tasks."
They envisioned that LifeLog technology "could result in far more
effective computer assistants for war fighters and commanders
because the computer assistant can access the user's past experiences . . . and result in much more efficient computerized training
systems."

We were excited after the stimulating brainstorming session around our conference table, and the project seemed to be building momentum. Was the U.S. Department of Defense, one of the world's largest organizations, going to be leading the way to the age of Total Recall? We didn't realize that LifeLog was headed into the middle of a political minefield.

In June 2003, William Safire wrote a column for *The New York Times* about LifeLog that put the fear of Big Brother into the reader's heart:

> And in the basement of the Pentagon, LifeLog's Dr. Gage and his PAL, the totally aware Admiral Poindexter, would be dumping all this "voluntary" data into a national memory bank, which would have undeniable recall of everything you would just as soon forget.

Although Safire seemed to finish the piece with tongue in cheek, invoking "Ned Ludd, who in 1799 famously destroyed two nefarious machines knitting hosiery," the powerful image of Admiral Poindexter (a key figure in the Iran-Contra scandal) in the basement spying on the population was enough to send the political class into hysterics.

Poindexter had also been the face man for another DARPA initiative called TIA, for Total Information Awareness, which was unveiled in the months following the terrorist attacks of September 11, 2001. The aim of TIA was to create a centralized überdatabase incorporating every electronic record, transaction, communication, file, and footprint the government could lay its hands on about every person and organization in the nation. They would then sift through this megadossier with data-mining software in search of patterns that could identify terrorist plottings.

Safire had blown the whistle on TIA as well, in an earlier column from November 2002. There was enough public outcry over
the possible abuses of TIA for it be officially scrapped a few months
later.

The stink over LifeLog seemed to rest on the fear-driven belief
that it amounted to the same thing as TIA. But there was nothing
about LifeLog that would have required people to entrust all their
personal data to a central server farm in the bowels of the National
Security Agency. There was nothing about LifeLog that even im-
plied people would be required to do lifelogging at all. This effort
was aimed at helping the individual soldier or officer in a state of
information overload.

I keep my nose out of partisan politics. I guess that made me
naïve enough to imagine someone would just explain the truth of
the situation (Safire hadn't even spoken to anyone at DARPA) and
sort things out. Instead, LifeLog was canceled. If I had cared more
about politics, I might have been outraged and suspicious that a lot
of political decisions were made based on juicy headlines rather
than common sense. In any event, I knew not to waste my outrage
on this, because I understood a little trick of technopolitics: Ideas
that run into trouble, especially good ones, are often officially
dropped only to be resurrected, recycled, and rebranded until they
gain acceptance. Technology does not give up or give in.

So LifeLog is dead; long live ASSIST! DARPA created the Advanced Soldier Sensor Information Systems Technology (ASSIST),
carefully explaining how it would help just soldiers. No one brought
up Admiral Poindexter this time, and the program went ahead.

A great example of the fine work done under the ASSIST um-
brella comes from the contextual computing group at Georgia
Tech, led by Thad Storter. They have shown the kind of ASSIST
daily impact could have for a soldier on patrol:

A platoon goes on a presence patrol in Iraq. Their goal is to be visible to provide a sense of security for Iraqi civilians, encourage goodwill, and look for signs of insurgents. Upon returning from their five-hour patrol, the platoon leader is debriefed by his intelligence officer.

"Anything unusual today?" inquires the intelligence officer.

"Pretty calm except that the children were acting strangely," he replies.

"How do you mean?"

"It's kinda hard to describe. . . ."

In Iraq, today's soldiers are fighting an insurgency that uses civilians for cover. According to the soldiers we interviewed, the most common point of contact with the enemy is the Improvised Explosive Devices (IEDs) used against vehicles and troops. Soldiers are forced to uncover the enemy through everyday patrols and intelligence gathering. However, the soldiers are facing an information shortage; they are not equipped to gather this type of everyday intelligence. Soldiers also need a means to share information with intelligence officers and between patrols. Currently, this information is mostly conveyed orally or through images taken with the soldiers' personal digital cameras. Georgia Tech's Soldier Assist System (SAS) attempts to augment this process by automatically capturing a "blog" of a soldier's patrol and allowing him to rapidly select media from that patrol to share with his intelligence officer.

To understand the goal of SAS, let's revisit the above scenario. While on patrol, the patrol leader and each of his two squad leaders wear SAS capture hardware. Each system records high-resolution images from a head-mounted camera, two streams of audio (one from a close-talking microphone and one from a chest-mounted microphone to record ambient sound), location using the Global Positioning System, and the soldier's movement using accelerometers on the wrists, hip, thigh, chest, and weapon. During the patrol, the soldiers can also use their high-resolution manual camera to capture images they feel may be important later. Upon returning from patrol the platoon leader now has the information to answer the intelligence officer's questions:

"Pretty calm except that the children were acting strangely."

"How do you mean?"

"It's kinda hard to describe, but let me show you."

The platoon leader now looks at a map with his GPS path overlain. He selects the area around a local mosque where he met the children and scans for an appropriate image. Looking for images where the system indicates he was speaking, he quickly finds an image with the children's faces to the intelligence officer.

"You see, generally the children will come up to us along the road because they know we carry candy for them. But today they are here along the wall of the mosque," states the platoon leader.

As the debriefing continues, the intelligence officer sees a suspicious white pickup truck in the background of one of the platoon leader's images. While the platoon leader's blog does not have a good image of the truck and its environment, he uses SAS's automatic annotation system to select images from his squad leaders' blog where they "took a knee" to provide security while he was talking with the children. (Often, when monitoring the environment to provide security, the soldiers support themselves on one knee while maintaining a good field of view.) The platoon leader quickly finds a good image of the truck and shows it to the intelligence officer.

"I bet you the owner of that pickup truck was there just before you and was scouting the area for the insurgency. Let's record this license plate and give it to the next patrol to look for," says the intelligence officer, ending the debriefing.

The Georgia Tech team found that, for the soldiers, "there was no such thing as 'too much information' for presence patrols." However, with all the data their system could collect, they didn't want the soldiers spending many extra hours wading through it all to find relevant parts to report. So, they did some postprocessing to automatically detect and tag activities like raising a weapon, walking, running, crawling, standing, shaking hands, driving, opening a door, and so on. These are not intended to replace the intelligence, intuition, memory, and common sense of soldiers, but to complement and enhance them. By combining these automatic tags with time and location, a soldier can quickly find the desired parts to report on.

When a new soldier rotates into an assignment, these improved, data- and media-rich reports will help him get up to speed, and will provide a new kind of resource to draw on. They will allow him to take over the memories of the assignment as well as the assignment itself. For example, if the new soldier sees a suspicious pickup truck, he can look at images from his predecessors' reports to see if it is the same one. Intelligence officers can go back to extra footage not included in the report to check out other elements that may not have been considered important at the time— was that old man near the truck spotted in the vicinity the last time the truck was seen?

In addition to DARPA, I've also met with CIA contractors to discuss applications for Total Recall. After all, the station chief in Budapest needs to hand on his memories to his successor too. It's another kind of tour of duty, where memories of faces, locations, vehicles, and so on can make life-or-death differences.

Total Recall is also important for the intelligence analysts at their desks back in the States. An interesting project called Glass Box records everything an analyst does at his PC workstation, literally recording a video of what is on his screen at all times, as well as tracking e-mail, opening documents, keyboard and mouse activity, Web surfing, instant messaging, and copy/paste events. The analyst can also make notes by talking or typing. Glass Box can be used to evaluate what research tools are most valuable to analysts, and possibly to detect traits of star analysts that could be taught to others.

Total Recall isn't limited to helping soldiers and secret agents do their jobs and fill their assignments; Every line of work will benefit from Total Recall.

Jon Gilmore worked as a Sprint engineer for nine years, designing new cell sites in regions where coverage needed improvement. He drove all over northern California measuring signal strength. He also visited individual cell sites to improve their performance, adjusting their power levels or tweaking the direction of an antenna. On his last day at work, Jon handed over the key to his filing cabinet and a hard drive holding about fifty gigabytes of information.

"I used to travel with a compass, a GPS, and a digital camera," says Jon. "I would verify the exact location of the site—often they were situated a little differently than what was in the records—and check the direction each beam was pointed in. I'd take digital pictures of the site, and the views from it to show the surrounding terrain."

"I also made notes and took pictures about access to each site. For instance, we used to call one 'ankle-biter lane.' If you read my notes, you knew not to get out of your car between the first two gates, unless you wanted to be bitten by the little dog there."

The engineer who took over for Jon also took over his e-memories.

YOUR REPUTATION

The Total Recall revolution will enable you to be the kind of employee or entrepreneur or small businessperson who gets more things accomplished, is more trustworthy, and more creative. The better you use Total Recall technology, the better your professional reputation will be.

Productivity gains will come from understanding one's work habits better. With a detailed e-memory of what I do, my computer is my personal time-management consultant. I can look back over my activity logs and notice where I've spent too much time on low-priority projects, or took too little time at a key place, or burned up a surprising number of hours reading Internet news. Mary Czerwinski's lab at Microsoft Research has come up with some brilliant visualizations of time spent at the computer based on keyboard and mouse activity associated with each running application. Most people are horrified at how often they are interrupted and at the time expended on overhead in their typical day. In the future, not only will we glean insight from such post hoc visualizations, we will program our cyber consultants to send us alerts and real-time reports to keep our time management on track.

Where there is repetition, Total Recall can spot it and take some of the drudgery away. How many times do we fill in the same form with nearly the same information, or very similar forms? Already most Web browsers have some sort of autofill feature to help us fill in online forms, but this can go much further. I get a lot of e-mails from students applying for internships, and I respond to most of them with one of a handful of stock replies running the gamut from "Unfortunately we have no openings now" to "We would be delighted to consider you." Software will soon arrive that will detect an e-mail as an application, dig into my e-mail history, and present me with a list of boilerplate options. Then replying to the current applicant is a breeze; it can be as quick as simply typing her name over the original name after the opening "Dear," or it might involve typing in a de novo paragraph that's specifically pertinent to her application.

Of course, boilerplate can easily go too far. Lawyers are already well down this road, where cut-and-paste is leading to a bloat in legal writing. I have hundreds of e-mails from attorney friends consisting of six words that were actually typed followed by a page of dire warnings and disclaimers. There will be misuse as with any technology, and that won't help your reputation, but this branch of workplace technology has only begun to fulfill its enormous time-saving destiny. And to make certain employees look extremely productive.

In general, Total Recall in the workplace means we stop doing so much work for the computer and the computer does more work for us. The software is inexorably moving from simple searches of huge e-memory databases to tools that manage the information coming out of e-memory so that it is especially relevant to the task at hand. Not a filing cabinet, but a sort of personal assistant. For example, Bradley Rhodes of MIT and his colleagues developed the "Remembrance Agent," an experimental piece of software that monitors your typing and reminds you of relevant e-mails or documents. If you are composing an e-mail and type "Project Anvil," the agent will bring up e-mails and documents matching those words on the side of your screen, ready for you to click open. In the same vein, Xobni has software to assist you with your e-mail. When you select an e-mail, it shows you contact information about the person the e-mail is from, a list of your recent e-mails, and a list of files you have shared with each other. Software assistants put information at your fingertips before you even ask for it.

Beyond being more efficient at the workplace or worksite—wherever that may be—when you are asked professional questions, you will be able to give an answer based on fact, not blurry bio-memories. You will be more reliable.

I often receive "remember me?" e-mails followed by some set of "action items"—to which I draw a complete blank. There are scores of former colleagues and potential business partners from decades back who try to contact me every month. It's not unreasonable, from their point of view, to expect me to remember them—many of them I worked with closely. Fortunately, about twenty years after leaving DEC, I was allowed to get my old files. They included eleven years' worth of correspondence, including hundreds of e-mails (which we had used throughout the 1980s, more than a decade before e-mail went mainstream). Those old communications have proved invaluable for cuing recollection of people from my past. I'm sure these contacts now think of me as someone they can better rely on; first to remember them, and then to actually recall details of connections in our work history.

Sometimes I want to dig up peripheral people rather than someone already designated as a contact in my address book, and having everything saved usually makes this easy. I can search through old e-mails from a friend to find one that includes the name of his son.

Or I may just search through everything I have for a "who the hell is that?" name and end up glad I scanned the program of the workshop that includes her name.

You never know what will be helpful. Before I really got religion on the "more is better" gospel, I tried to talk Vicki out of scanning all my old high school yearbooks. "How could that be of any use?" I asked. Fortunately she ignored my protests, and lo and behold a few years later I received an e-mail from a Dr. Tom Hill, a successful entrepreneur turned corporate team-building consultant, asking me for some biographical information so that he could describe my work in Tom Hill's Friday Eaglezine. He identified himself as being a 1953 graduate from the same high school that I had graduated from in 1952, but his name didn't ring any bells for me. A search for "Tom Hill" in the yearbooks pulled back photos and descriptions of various activities that brought the high school memories flooding back. When we spoke, I was able to recall half-forgotten events and people we'd clearly known in common, which made for a pleasant conversation. I'm now part of his Eagle network.

The "who the hell" problem only gets worse with a person's age. With the advent of social networking sites such as LinkedIn, "remember me?" messages and invitations to join yet another group are constantly pouring in. By hanging on to three decades' worth of e-mails, business cards, meeting appointments, photos, and audio recordings and using MyLifeBits to group and interlink them, my contact management has reached a whole new level.

What you may scorn today may one day turn out to be practically useful. A key reason people leave established companies to form start-ups is to get away from the numerous and stultifying rules and procedures. Ironically, one of the first things they miss is some of that important red tape. In 1988 I was head of engineering at a start-up, and we realized we needed a product release process. Thankfully, we saved a lot of time when an employee turned up a copy of the release process document from Sun Microsystems. It was equivalent to a process we had used at DEC, so it didn't take much work to turn it into something I was happy with. I continue to receive requests from others in the same boat for items like DEC's engineering handbooks.

As work experience becomes more of a scientific record and less of a befuddled bio-memory, your work time will become more creative. First, you simply won't have to argue about what happened anymore, and second, the interconnectivity of e-memory records will free you to make new associations.

THE E-MEMORY ENTERPRISE

At the same time Total Recall is changing your work experience, it will be changing it for everyone else. Certain organizations will make better use of it than others. Many of them will make office e-memories available to everyone who has a similar job within the company or institution. It's safe to predict that the same trends in technology that will lead individuals to use e-memories will also lead organizations to make the most of the new technology. Many will soon be keeping everything from internal meetings, e-mails, and memos to external-facing activities like sales, customer support, and purchasing. And making it searchable, usable information.

Time management, knowledge mining, trend discovery, context-sensitive reminding, and other computer assistance will also be applied to the institution's broader memory. SRI, the institute where Doug Engelbart invented the computer mouse in 1964, is managing an extensive DARPA-funded research program called the Cognitive Assistant that Learns and Organizes (CALO). CALO presumes a corporate digital memory, and rather than requiring users to learn all of its ins and outs, it learns from its users and from the material their organization produces. It learns about people's needs, routines, and expectations to become a real assistant that can be proactive. Software companies such as DEVON are already pursuing directions like this with products like DEVONthink, which aims to detect complex and subtle connections between documents that can stimulate new ideas.

Another example of progress in this area is a collaboration between MIT and Hewlett-Packard known as DSpace. Launched in 2002, DSpace is an open-source software package designed to accommodate an institution's entire body of records, resources, and output, including books, aging microfilm, administrative records, audio recordings of classroom lectures, video recordings of speeches and events, scientific research data, published papers, student theses, 3-D models and scans of objects, and any other kind of digital information. The software includes a search engine and the ability for users to tag information in order to create useful trails and associations not present in the original data sets. DSpace has been adopted by hundreds of universities and research centers worldwide.

I expect Sprint will make Jon Gilmore's memories about a cell tower available to any engineer who works on the tower, not just his successor. Likewise, repair divisions, like that of Xerox, use a common knowledge base to share diagnosis and repair information; all technicians inherit the memory of the one who first solves a particular problem, and get notes and tips for each model and type of breakdown. So when a technician first encounters error message #104 on model C900, the stories of a couple of previous repairs are instantly available, along with a tip to check for foreign material in the paper feeder before assuming the module needs replacement.

Customer service would sure be a lot better with a divisional memory. I can't wait for my mobile phone provider to get onboard with this, because I'm sick of endlessly recounting the same story during a series of calls to different representatives, not to mention battling their skepticism toward my claims of what previous representatives have advised me. The next step is to make the memories accessible across the whole company. I want the wireless Internet troubleshooter I speak with to have access to, say, the billing memory to connect the dots when billing is really the issue. I've spent enough hours of my life on hold being passed around between specialists from that company, because the right hand doesn't know what the left has been doing.

The march toward institutional e-memory has begun because the adoption of digital storage and communications makes recording and retrieval just too cheap and advantageous to pass up—especially in a competitive corporate environment. Keeping e-mails, instant-message chat sessions, and transaction records is obvious. With all bits of communication becoming virtually interchangeable you can generate voice from an instant message, or use voice-to-text to search for e-mail. Whenever I call my bank or insurance company, the first words from the corporate mouth are either "This call is being recorded for training purposes" or "This call is being recorded for your protection." In a call involving stock purchase or sale, it is not uncommon for two agents to be involved and to verify the correctness of a verbal transaction. In companies such as Hartford Insurance, all sales calls are recorded along with all the interactions to the various databases that make up a call, making it possible to virtually recreate the full interaction and its associated data. In the next few years, we will see calls like these being converted to text so that when the customer calls again, the company representative will have the transcripts up on-screen during the call.

In addition to all communications, inventory is going to be tracked in greater and greater detail. Just as you have come to expect Federal Express to know the history of the packages you ship, construction companies will know the history of every sheet of drywall they use, and sporting-good manufacturers will know about every baseball glove they make from the time it is "born" to the point of sale.

The same trend will apply to individual tools and pieces of equipment. Each item will carry its own unique network identity so its usage can be logged and tracked, including who used it, where, and when. For instance, I presume that banks already know which ATMs are being used, along with when and by whom, as well as who is servicing them and when. This kind of fine-grained logging and tracking will expand to virtually everything: a dentist's drill, a gasoline pump, an inventory scanner in a warehouse, an espresso machine at Starbucks. It will even include meeting rooms, which may have audio recording in addition to just the log. The traffic of people through public spaces will be studied and learned from.

I like to think about how Total Recall might have impacted my father's business, Bell Electric. Dad started the company in 1933, and ran it until he had a heart attack in 1985. Bell Electric sold, installed, and repaired electrical appliances and equipment. A large NCR cash register printed a record of all the transactions that went through it. The main ledger tracked purchases, sales, electrical installations, and repairs. Bills were sent out monthly. Every year, he would close the shop for a day to take inventory.

An e-memory for Bell Electric would subsume Dad's ledgers and various record books. All of his personal knowledge of his customers would be in customer-relations management tools. Instead of flipping through the W. W. Grainger catalog of electrical equipment, and dealing with paper orders, invoices, and receipts, Dad would order from their Web site and save all the transactions electronically. Having to order a weird replacement part for a pump would be much easier the second time around because he would instantly recall the first order. Inventory would be known continuously, and the complete history of a certain customer could be recalled in a moment. The flow of customers to the store would be analyzed to recommend the best hours to keep and when to take holidays with a minimum impact on the bottom line. Dad would know his profit by employee, by customer, and by type of job. When one employee followed up on the work of another, he'd have the former's e-memories, and know there was a second circuit panel in the family room, that the wires had nonstandard colors in the kitchen, and not to call the customer after nine P.M. Dad would learn to manage his own time better, based on his lifelog.

The center of every organization, large or small, will be its institutional e-memory. E-memory will be the heart of customer service, human resource management, strategic planning, inventory, shipping, finance, payroll—everything. And with data mining, every aspect of operations can be analyzed and improvements formulated.

The only thing in the way of an institution's e-memory is their legal department, who often mandate the deletion of records such as e-mail to limit liability. It remains to be seen whether lawsuit settlements can continue their mind-boggling rise to stay ahead of the value of corporate e-memory in the Total Recall era. After all, these records will also include things like alibis against some charges, and proof of an idea formed prior to a competitor's patent. Furthermore, there is usually someone, somewhere, who has kept a copy and thwarted the lawyers' intentions. It never pays to take lawsuits lightly, but I don't see how corporate e-memory destruction policies can continue.

THE FAMILY ENTERPRISE

Everything I've said about increased productivity at work could also be applied to your personal life. You might get a lot out of understanding how you spend your leisure time, and of course wonder, "Who the hell is that?" at home just as much in the office. I know I need a cyber assistant for my personal life because I already need Vicki's help as a personal assistant with things like travel plans for my family vacation.

And while the impact of Total Recall on your professional life will be far-reaching, the home office is where you will feel the personal payoff. The family is an enterprise, much like any business, with financial and legal matters, schedules and plans, and records to maintain. You need to keep track of family members and property such as cars, homes, and investments.

I have more than a hundred unique kinds of items in my e-memory that are part of my virtual home office. There are legal documents like wills, deeds, licenses, and birth certificates. There are all kinds of financial and tax records. A home loan can consist of several hundred pages and dozens of documents with signatures. Your home itself may have wiring diagrams and blueprints. Cars have loans and maintenance records. And every appliance has its warrantee and manual.

Having all this personal information at your fingertips really helps. Recently, I had to fill out a form for the Australian government enumerating all the countries I have visited over my lifetime, including the time and duration of the visit. That would have been daunting in the old days, but with e-memories it was no problem. Jim Gemmell has saved time and made arrangements more quickly on numerous occasions simply by having scans of his children's birth certificates handy for sports teams that are constantly demanding them. I can't tell you how many times I've been on a trip and can't remember how to use some feature of my camera. I refuse to carry an instruction manual that is bigger than the camera, but now that my camera's instruction manual is in my e-memory, it's never a problem anymore.

E-memory will be of great value to all kinds of organizations. It will be at the heart of businesses like law firms, software companies, hospitals, banks, retail stores, electricians, winemakers, and airports. It will be employed by nonprofit organizations of all kinds, including homeowners' associations, school boards, lodges, churches, and hobby clubs. From the boardrooms of gigantic corporations down to the kitchen table of a small family, Total Recall will help get things done.

CHAPTER 5

HEALTH

"Let's have a look," said the doctor.

I parted the baby-blue hospital gown to expose my chest, which was dappled with faint red blotches. The doctor peered at them appraisingly.

"Yes . . . well . . . they look a bit better, don't they?" he said. "I think we can get you out of here by Saturday."

It was the middle of August 2007, about a month after my second double-bypass surgery. I was in the hospital because those faint blotches might indicate a complication that could lead to another operation. I didn't say anything, but I knew for a fact that the blotches had not changed. I knew this because I had been taking pictures of them daily with my digital camera and comparing them side by side on my PC.

The reason I kept mum was my desperation to be discharged by Saturday so that I could celebrate my birthday at home. I probably should have told him the truth, but I was miserable after almost a month of hospitalization and reasoned that I would start running a temperature if the situation became serious. The doctor, relying on his memory, discharged me. I got my wish for a birthday party at home, and the blotches eventually went away.

This episode illustrates how often professional health care strays from quantitative analysis. "How long have you had the fever?" asks the nurse, and I struggle to pick a likely time. "Do you recall what you ate before the migraine?" asks the physician, and I realize I have absolutely no clue. I was supposed to be noticing? Then there's that pseudoquantitative classic: "On a scale of one to ten, how much pain do you feel?" Doctors hear plenty of vague and qualitative complaints: "I've been feeling run-down for weeks" or "I get these pains sometimes." What a difference it would make if patients could follow their complaints with, "Here's a graph of my temperature every hour for the past two weeks," or "Here's a time line of everything I've eaten in the last month, with times of migraines noted," or, in my case, "Here are twenty photos of my rash, taken daily."

In the Total Recall world, health records will be transformed into minutely detailed chronicles of vital signs, behavior, diet, and exercise along with physicians' diagnoses, prescriptions, advice, and test results. Your e-memory software will make managing this total health record easy, and you will be healthier.

HEALTH E-MEMORIES

It was recently reported that a university study of more than forty hospitals and 160,000 patients showed that "when health information technologies replace paper forms and handwritten notes, both hospitals and patients benefit." Neil R. Powe, M.D. from the Department of Medicine at Johns Hopkins University School of Medicine and director of the Welch Center for Prevention, Epidemiology and Clinical Research was the lead author on the paper announcing the findings. He said, "If these results were to hold for all hospitals in the United States, computerizing notes and records might have the potential to save a hundred thousand lives annually."

Good information is central to good health care, and the old-fashioned paper-based system is inadequate. Most hospitals have not caught up with the efficiencies of our digital world. Many laboratory tests are performed needlessly because of missing paperwork. One study found that while the patient's chart was available 95 percent of the time, 81 percent of return visits were plagued by missing information. RAND has estimated that the U.S. health-care system could save an estimated $77 billion each year from the improved efficiency of electronic health records. Health and safety improvements double that figure.

The paper-based system is not just inefficient; it can be dangerous. In American private-sector hospitals and nursing homes, as many as one in five medications are given in error, harming at least one and a half million people every year, with 7 percent of those errors being potentially life-threatening.

Compounding the issue is a projected increase of chronic illness. We are in for an explosion of chronic ailments as the Baby Boomer generation passes into seniorhood through the 2010s and -20s. Eighty-eight percent of seniors have chronic conditions that require ongoing management and become increasingly expensive the longer they are left untreated. As the baby boom becomes the senior boom, our health record problems will multiply.

Thankfully, paper-based health systems are on their way to extinction. Health-care providers around the world are moving to electronic health records, keeping an e-memory of your medical records instead of paper. Being digital, they can be easily accessed, copied, or updated from anywhere within an institutional intranet or, in some cases, via the World Wide Web. The institution that maintains your records uses them to log every health-related event and transaction relating to you as a patient, from treatment records to consent forms to insurance billing to test results.

The United States Veterans Administration, which cares for the health of the nation's ex-military personnel, has adopted electronic medical records and other computerized systems with fabulous results. It has almost eliminated prescription errors and the need to duplicate lab tests. Many private American institutions are also implementing electronic health records (EHRs). Kaiser Permanente is aiming to deploy the nation's largest electronic medical records system by 2010, covering 8.4 million members, 431 medical offices, and 32 hospitals in 7 states. Hoping to shave down high costs of providing insurance for their workforces, a consortium of major companies, including Intel, Wal-Mart, and AT&T, are working to provide electronic health records for their employees.

State-run medical systems can dictate EHR adoption, and even the central storage of all information. The British National Health Service is mandating a move away from paper, with a central EHR system dubbed "the spine." The European Union adopted an e-Health action plan in 2004, which includes e-prescriptions, e-referrals, and teleconsultations.

But while there has been considerable progress toward EHRs, a lot of work remains. The San Diego County Medical Society Foundation believed in 2003 that they were only a year away from a regional information network but, as of 2007, still remain "years away from full scale EHR adoption," according to *The San Diego Union-Tribune*. In Santa Barbara County, a nonprofit organization was started in 1999 to set up EHRs in hospitals and doctors' offices.

However, when a $10-million grant ran out in 2006, the effort was abandoned.

One significant hurdle is demonstrating the value of EHRs to physicians. Fewer than a third of American doctors currently use electronic records. They rightly worry that promised savings may accrue in other parts of the health-care system, leaving them with no reward for their investment. And without an agreed-upon or even a de facto national standard, they may have to scrap or retool their electronic health records if they end up betting on the wrong format.

Worse, when different systems use different standards, it is difficult, and even dangerous, to share information. For example, the code "DPT" may stand for Demerol-Phenergan-Thorazine in one system but the Diptheria-Pertussis-Tetanus vaccine in another. Imagine being admitted to an emergency room and being dosed with a drug you're allergic to because they misinterpreted the code in your record. And yet there are some twenty or thirty commercial electronic health records on the market, most of which use proprietary data formats. Many health-care providers have watched cagily from the sidelines, waiting for a standard to be set. Fortunately, it appears that waiting will soon end, as recent years have seen spreading standardization as well as software coming to market that translates between standards.

At the same time, some health professionals are exploiting unorthodox tools for organizing and sharing medical information right now. In a recent bit of serendipity, radiologists realized that they could use iTunes, Macintosh's popular music-management program, to manage and organize PDF files just as easily as music files. For example, radiologists generally save medical reports as PDF files in folders on their PCs. If they want to compare images from several articles, they have to pull each one out "by hand," so to speak, and lay them side by side. But iTunes can search, describe, and rate PDFs just like music files. The physicians can search and sort PDFs according to any desired criteria and publish their "playlists" for other doctors to see, just like music lovers do with their MP3 songs. Bring up all X-rays of fractured tibias within the past year. Show me all herniated discs in people over age seventy-five stored in my PDF files.

Health-care providers employing PHRs will have decreased costs and better service. This will bring them more business, and force their competitors to adopt PHRs as well. All signs indicate we have already reached the critical mass needed to see personal health records become universal.

OWNING YOUR HEALTH MEMORIES

The typical American can access her bank account from ATMs around the world, check her e-mail from anywhere there's a Wi-Fi signal, and go online to see charts breaking down exactly how energy is used in her home. But if she wants to look at her own health records, she's out of luck. The little she can view online is incomplete and spread among many sites, and in any case, most of the relevant information exists only on hand-scribbled forms and printouts stuffed into a dozen manila file folders in a dozen far-flung offices around the country.

No one institution has the full picture of her health. The hospital has one piece of the puzzle, the specialist another, and the family doctor yet another. Her insurer knows everything that has been billed but lacks most of the details, and it has no idea about her out-of-pocket visit to the naturopath, the chiropractor, and the cosmetic surgeon. Factor in the dentist, the pharmacist, the gynecologist, the dermatologist, and the therapist. Her current general practitioner keeps his own file on her, but everything he knows about her prior medical history consists of whatever she happened to remember (or misremember) on her first visit.

American health care has been fragmenting. Driven partly by technological advances and partly by the bottom-line focus of our system, hospital alternatives are popping up all over the place. Step right up and get your full-body MRI scan in the shopping mall. Measure your blood pressure at the pharmacy counter. Drive in and get your cholesterol measured. Come into the workplace clinic for free advice. Deliver your baby at home using this highly recommended midwife.

Clearly, only the individual is in a position to be at the hub of all his health information. Only you have the right to all your health e-memories. Indeed, you may log some aspect of your health that no one else has a need or right to see. You must take ownership of your health memories.

Seeing the need, software makers have already taken action. Quicken Health can track the state of all the financial transactions associated with health treatments. This is essential for billing and insurance information for the 88 percent of us who have some chronic health problem as we pass sixty, since every medical encounter generates many pieces of paper that we are likely to have to deal with.

Microsoft HealthVault is a free service that promises to warehouse and safeguard your personal medical information that can ultimately be shared with health providers. Google Health is another such utility. These services allow you to upload and manage your own health and wellness information and to authorize third parties to "blindly" upload data to your record, without being able to access the other information therein. The third party can be a person, such as your doctor, or an organization, like an insurance company, or a private clinic, or even a third-party software application. For example, you can have your weight and blood pressure stored to your HealthVault courtesy of software developed by the American Heart Association.

I think of HealthVault for health much in the same way as Quicken or Money for finances. My financial transactions come in different types, and so do my health records. Just as banks, credit cards, and brokers are combined under one database, records from each physician are aggregated into one database. Thus you become the keeper of all your records.

Someday, collecting all your health e-memories will be a snap; today it is a challenge. When I decided to pull together my medical information in 2001, my health records were strewn across four states. I collected my medical files from general-practitioner internists; heart, eye, and other specialists; dentists; several hospitals and clinics; and the half-dozen insurance companies that have covered me over the years. I ended up with more than a thousand pages.

None of the material was in digital form, even though much of it once lived on a computer. Some came by fax because that was deemed secure while e-mail wasn't. Some was on large negative film, for example X-rays and MRIs. I scanned all the material into MyLifeBits. It included consultation records, doctor communications, surgical reports, immunization schedules and records, pharmacy orders, optical prescriptions, gum depth measurements, explanations of benefits, lab test results, receipts, electrocardiograms, pacemaker data, and echo stress tests. Several nuclear stress tests measuring heart blood flow on film and VHS tape had been discarded. Thus the only hard data about my heart over a long time period were lost. Luckily, a 1995 angiogram film of my heart's vessel had been retained and was useful for the surgeon on the second bypass.

I had detailed physician statements about my heart, because my mother saved them, going back to a family physician's letter observing a murmur when I was eight years old. I was immediately confined to my bed for that summer, followed by a visit to the Mayo Clinic.

In 1956, when I was in my early twenties, a cardiologist at the Massachusetts General Hospital reevaluated my situation. His prognosis: Go live a normal life without competitive sports. All through the sixties and seventies my general practitioner assured me that a cholesterol level of 230 was normal. Maybe so, but "normal" turned out to mean "bad" for me when I had a cardiac arrest on February 27, 1983, in Snowmass, Colorado. My friend Bob Puffer gave me CPR and saved my life. I spent the next ten days in a coma in Grand Junction. On recovery I was flown by helicopter directly to a double-bypass operation in Denver. There was some concern about impact to my brain, so the doctor asked me some post-op questions.

"Who is the president?" he asked.

"It doesn't matter," I replied.

My friends chuckled. Everyone knew I was right back to normal.

There's nothing like a brush with death to motivate you to make changes. My heart attack lit a fire under me to start taking my health much more seriously, which included my record keeping and self-monitoring. It was one of the factors that fueled my interest in starting up MyLifeBits. I've been keeping my own records since then.

HEALTH LIFELOGGING

Many machines in hospitals spew out paper, or even worse, no record at all, just ephemeral blips on a screen. A whole day of data collection may be summarized in the record as just "normal" or "elevated." Even in nominally stable vital-sign data there can be patterns or brief events that could have relevance to treatment or diagnosis. Why aren't more health-monitoring devices integrated into the information network, with their results saved?

The technology has not been there for us, but it soon will be. Abundant storage means we can afford a complete health lifelog. There is already a trend to collect and keep more data from existing medical equipment. Even more exciting, a whole host of new biosensors are coming on the scene that will expand our knowledge almost beyond imagination.

Biometric sensors are moving from the clinic to your home—or wherever you are. Diabetics measure their own blood-glucose at home from a pinprick of blood. Asthmatics breathe into palm-sized devices that measure their airflow. Finger clips can measure pulse, blood and tissue oxygen, blood sugar, the proportion of red blood cells (hematocrit), and tissue acidity (pH.)

People with sleep apnea can now be monitored, at home, in their own beds, with a variety of devices. Apnea, or brief cessation of breathing during sleep, leads to heart disease and other chronic problems. A clinical sleep apnea test involves spending the night in the clinic, covered in wires, constantly watched over by staff via cameras above the bed. In contrast, in-home methods are being developed that use finger (or toe) clips and sensing wristwatches, with microphones and infrared cameras tracking movement but protecting privacy by not recording actual sound or images.

If you belong to a gym, you've probably seen fitness equipment that tells you your heart rate and how many calories you've burned. Get ready for many more fitness sensors. Nike and Apple have partnered to develop sensors in the soles of shoes that transmit to your iPod. During your run, the iPod displays your pace, distance, time, and calories burned. After the run, the workout information can be uploaded to a workout e-memories Web site. Some gyms also have cardio equipment such as treadmills or stair steppers that can send data to the iPod to record workout data.

I bought a BodyBugg, which you wear on an arm strap against your bicep. It measures temperature, heat flux, galvanic skin response, and acceleration. Heat flux indicates how many calories I am burning. Galvanic skin resistance measures tiny fluctuations in my sweating; it's one of the main physiological signals used in polygraph "lie detector" tests, and indicates psychophysiological parameters such as stress, anxiety, arousal, and surprise. From this data, I get a report that includes my calories burned due to activity, calories burned during rest, physical activity levels (in METs) and durations, sleep duration, and sleep efficiency.

I'm also intrigued by experimenting with smartphones that come equipped with accelerometers and GPS units. GPS and wireless signals from cell phone towers and Wi-Fi hubs can track my location and movement, which is a highly relevant component of health data. Location data can tell me how much "bonus" exercise I net in my daily comings and goings and it can show me how close I'm coming to my recommended "ten thousand steps per day" plan.

People with congestive heart failure need to track their weight gain, which may signal an increase of fluid retention due to poor circulation. They are prescribed diuretics to shed the fluid. But if their weight gain is due to increased muscle mass, from working out at the gym as the doctor ordered, diuretics are the last thing they need. Thus, scales have been developed with a handle that passes an imperceptibly mild electric current from hand to foot. Changes in electric conductivity indicate the nature of the weight gain—fluid or muscle—so patients can discern whether their weight gain is harmful or beneficial.

Another way to get health data from me is to build sensors into my clothing. Researchers at Dublin City University, Ireland, are working on fabrics that can be made into shirts that track your breathing, or in your shoes to track your steps. They also have a treated fabric that, together with a small LED light and sensor, can be used to detect the PH level of your sweat—an indicator of dehydration. Their fabrics may also be used to detect your posture (another way to track posture is with cameras, such as Alexandro Jaimes and Jianyi Liu used to warn the user in front of a PC when an unhealthy posture has been held for too long).

In the future, the most amazing sensors will be implanted inside your body. Those of us with chronic ailments like diabetes or heart conditions are likely to have implanted sensors that wirelessly transmit their knowledge to another device outside the body, such as a cell phone or personal digital assistant. These sensors will not only stream our vital-sign readings to our personal health record but will continuously monitor them for troublesome or telltale patterns. Depending on the severity or risk, they will e-mail us alerts to follow up with our doctors as soon as possible, or immediately connect us to our doctor's office, or even autodial 911 and send for an ambulance. Cardiac devices such as my pacemaker are already being equipped for wireless communication. They pass on values undetectable outside the body, including electrical activity, intra-ventricular pressures, blood flow, and ejection fractions.

All of your biosensors will communicate to have their data become part of your lifelog. Manufactures like Philips already sell a line of such devices for the home that wirelessly transmit to a hub that then can forward information to a health provider. I recently bought a Bluetooth-enabled bathroom scale that automatically sends my weight to my e-memory (in this case, Health-Vault) where I can chart my weight-loss progress, or lack thereof, over time.

In addition to sensors, the reduced cost and increased convenience of some lab work will expand our health lifelog. For example, comprehensive blood sampling is becoming cheap and possible to do without large volumes of blood, enabling biomarker testing that can help identify the onset of many conditions, including cancer, cardiovascular risk, and autoimmune diseases.

In November 2008 I accepted an invitation, as a Microsoft employee, to participate in a landmark digital health research study called the Scripps Genomic Health Initiative. They asked me to send a DNA sample—via spit in a bottle—to San Diego for gene sequencing. The study is aimed at understanding if people will be motivated to make positive lifestyle changes such as exercising, eating healthy, and quitting smoking after receiving their personal genomics test report.

I believe more testing will become routine, and we will log all the details of each test, not just their summary results.

You don't have to be a multiple heart-attack victim to see how much life quality, and even years of extra life, will be salvaged by recording how well your body is functioning. Then again, even multiple heart attacks can't get me over my loathing for dealing with batteries. The hassle factor of changing batteries all the time is just too much. Here's a word to the wise for all the miniature-device makers out there, whether it's cell phones or cameras or medical devices that you wear. At minimum, a standard USB cable should both charge and upload data at the same time. And, yes,

I said standard—enough of your special thirty-dollar cables and cradles. The ultimate solution is that I should just be able to toss my devices on a universal charging table that recharges them wirelessly through induction, while they upload their data wirelessly to a browser-accessible place in the cloud in a format a middle-school student could read.

My loathing of batteries keeps me from using my BodyBugg most of the time. But really, I should know from experience not to skip any biomonitoring. I used to wear a Polar strap around my chest when I bicycled so I could get just the right amount of exercise. Regrettably, I didn't wear it during a biking trip in the hilly region around the French Riviera in April 1996. I was not fully recovered from the flu, but didn't realize that would matter. After a day of biking and a lovely dinner, I awoke with a severe pain that I attributed to "heartburn." In fact, it was a heart attack that needed immediate attention. I had overexerted myself. By not wearing the Polar strap, I'd left myself in the dark. The consequence—a blowout of part of my 1983 double bypass—was revealed in a stress test several months later.

We are only beginning to imagine what can be tracked with implanted sensors. Combined with wearable sensors, in-home sensors, and ever cheaper, more accessible laboratory tests, a whole new universe of quantitative health is dawning.

GETTING HEALTHIER

This chapter opened with a story about my doctor relying on his memory to gauge my improvement. But doctors aren't the only ones with imperfect memories and filtered perception. Patients don't hear half of what they are told by their physicians, even when especially important information is discussed. I believe it would be very useful to record these sessions with our physicians to aid our memories in these times of stress.

Just remembering to take your pills is a problem. People usually don't forget painkillers, because pain is a potent motivator, and its own reminder. But overall, patients only take about half the medications prescribed to them. Have you ever taken your medications until you start to feel better, then quit taking them days or weeks before completing their course? Have you ever left your doctor's office all revved and psyched up to eat a healthier diet, start exercising, or floss daily only to find the new habit lasts but a week?

The good news is that quantitative health can be a shot in the arm for motivating you to take care of yourself.

You are more likely to take your medications if you can see a chart that confirms the value of the medicine. What if you had a chart on your personal Web page showing your cholesterol going down week by week? What if you got quarterly imaging scans of your arteries so you could watch as plaque dissolves from improvements in your diet and exercise regime? What if you could observe your heart and lungs in a series of images, or your vital statistics trending up or down on a chart? Quantitative health data is already being used to motivate patients, and the practice will only expand.

Your personal health record combined with biometric lifelogging can be integrated with your complete e-memory. As with everything else in Total Recall, the more integration, the better. You should be able to find all sorts of informative correlations between your health data and seemingly miscellaneous other facets of your life.

Imagine if your e-memory presented you a chart revealing a high correlation between your entertaining a certain difficult rela-

tive and a trifecta of weight gain, poor sleep, and self-imposed social isolation for a week following each of his or her visits. You'd
think, *Wow, I had no idea it was so bad*. Knowing this, and assuming
you couldn't avoid the person by dint of filial duty, you might at the
least take pains to manage your exercise, eating, and socializing in
the wake of those visits.

Or what if you could see in plain-as-day graphical form that 80
percent of the time when you order a red-meat entree at a restaurant, you become on average 30 percent less productive at work
and half as likely to go to the gym the day after? Armed with such
knowledge, you might cut back on red meat or redouble your vigi-
lance against lame rationalizations for not exercising.

Or what if your e-memory gave you a chart showing a connection between moderate exercise and improved sleep? That will be
more effective than generic advice or nagging to motivate you to
get out and walk more because your own (digitized) experience
will be telling you this.

In July 2008, a year after my second bypass, I felt some slight
angina pains while walking to work in the mornings. This could
mean that my arteries were getting clogged again, which would be
extremely bad news. So I decided to log my weight, diet, exercise,
and heart rate to see if I could glean any patterns. Would eating a
dip or two of ice cream have repercussions? Were the tingles in my
chest related to times or distance that I walked?

I carried a small GPS tracker on my way to work, measuring
the distance walked and elevations encountered—remember, I live
in San Francisco. I wore a Polar heart monitor and pedometer for a
record of my heart rate and footsteps. The monitor makes suggestions for my weekly workouts based on height, activity level, and
my weight. With GPS and a pedometer, it can download workouts
from the Internet and keep track of all my exercise.

I used my BodyBugg to look up how many calories I'd burned in the past two days, or since breakfast, or in the past ninety minutes. I compared my average burn rates on weekends versus workdays. In the evenings I checked whether I'd burned more calories than I'd consumed that day, which I could use to justify a little ice cream.

I even made a point of wearing my SenseCam during meals to see if it could give me pictorial reminders of the food I ate, as opposed to what I remembered eating. How many shrimp did I really consume? What was on that snack table in the coffee room?

I learned that the angina was related to my food intake. By increasing my exercise and reducing fat even more, I was pain-free again after six weeks. Ice cream and even low-fat cheese are verboten, just as I had given up butter and foie gras in '83 after my first bypass. I am becoming convinced that one can eliminate plaque with a regimen of diet and exercise, just as physicians like Dean Ornish claim. If only there were a way to easily measure that buildup.

The benefits of health lifelogging are irresistible: increased self-reflection and self-knowledge, less room for denial or half-conscious fudging on your diet-tracking or time spent at the gym, improving your health habits, helping you cope with or cut down on stress by identifying its causes, alerting you when you get swept away by negative passions, and saving your life by identifying incipient strokes, heart attacks, panic attacks, and other acute episodes.

ENTER THE PROACTIVE ADVISOR

Farther down this road, we will see the advent of the e-Nurse. Timothy Bickmore of Northeastern University has developed a virtual health coach called Laura. Laura is a computer-generated character who nods and raises her eyebrows as she engages in conversations with patients.

Laura has increased the physical activity of elderly patients by 100 percent, and is used to help schizophrenics stick to their medications. While e-Nurses can never be as good as real nurses in many ways, it turns out that patients are sometimes more comfortable asking questions of a virtual health provider, rather than taking up the time or asking potentially embarrassing questions of a real person. Besides communicating with patients using an animated character, e-Nursing can take place via text messaging, cell phones, chat sessions, or any electronically mediated form of communication.

With all this in mind, you should expect to hear stories like these soon:

Sara gets an e-mail from her family's e-Nurse, recommending that she take her son, Alex, in for a checkup. The e-Nurse has noticed that Alex's weight, wirelessly submitted from his bathroom scale each day, is tracking lower than expected. The e-Nurse suggests that this could be a side effect of Alex's asthma medication. Alex also puffs into a meter each day, which has likewise been wirelessly transmitting his breathing flow, and his breathing has been excellent for several months. The doctor reviews the situation and suggests taking Alex off his medication.

Her sister, Gwen, has been sharing her moods and sense of well-being with her e-Nurse. The e-Nurse begins asking questions about her diet each day. After a few months, the e-Nurse suggests a link to wheat. Gwen is tested and discovers that she does indeed have a wheat allergy. Going off wheat radically improves her moods and energy levels.

Their father, John, lies down in his bed, and a wireless unit underneath the mattress communicates with his pacemaker, downloading the story of his heart for the day. Almost every month, his medication is slightly adjusted based on pacemaker data. Several times, a trend of his weight combined with heart activity leads to messages from his e-Nurse. The e-Nurse remarks that these episodes seem to follow times the RFID sensor in the fridge has tracked chocolate ice-cream purchases. John has believed he could get away with a "little bit" of his favorite dessert, and is chagrined to learn he cannot.

Our health care has been built on limited, spotty data. It reminds me of the guy building a house mostly by "eyeballing" it, with only rare use of his tape measure, level, or square. Health care with Total Recall is like a house that is built right.

A HEALTHIER WORLD

Tremendous collective benefit will be gained from pooling personal health data. A population's worth of personal health records will be invaluable in large epidemiological studies.

Location information can be essential in discovering health-related environmental factors. The great breakthrough in understanding cholera in the mid-nineteenth century came from associating it with certain water supplies. Who knows what great strides in epidemiology might be made based on correlations between health and location data? Add into the mix diet, exercise habits, social patterns, and literally hundreds of other dimensions of how people live, and you are assuredly looking at a tool for improving public health in the same league with the greatest modern-health achievements including germ theory, immunization, and antibiotics (not to mention plumbing).

You will soon be seeing massive health studies as never before, but they will no longer be high-maintenance, expensive endeav-

ors. Instead, medical scientists will be able to ask simply for ano-
nymized elements of people's e-memories. Analyzing such data
across thousands or even millions of individuals, they will be able
to study correlations of a sort previously confined to speculation.

Some applications won't be anything like what researchers
have thought of as epidemiological study. Google is able to track
the spread of the flu by noting when people enter words like *flu
symptoms, aches, sore throat, cough,* and *fever* into the Google search
engine. This simple act multiplied across millions of keyboards
provides an early warning system for the spread of the flu about
ten days before the Centers for Disease Control and Prevention has
similar information collated from emergency rooms and health
departments around the country.

Total Recall promises a revolution in personal and public
health.

CHAPTER 6

LEARNING

When Deb Roy's son was fifteen months old, a video camera in the hallway ceiling documented his first tentative steps.

As the boy starts to totter toward him, Deb asks, "Can you do it?"

The toddler staggers on. Amazed at this new, upright world, he whispers, "Wow."

Deb watches in rapt attention. Then he, too, says softly, "Wow."

The Roy family has this wonderful Total Recall moment thanks to their hallway camera, but family memories are really only a fringe benefit. For Deb, recording is all about learning—learning about the acquisition of language, in particular.

Deb Roy is the director of the Cognitive Machines group at the MIT Media Lab. In 2005, Roy and his wife, who is also a professor, had a son and decided to make all three of them into guinea pigs in a self-run experiment. They wired up their house so that virtually everything their son would hear and see from birth to age three would be recorded. Their equipment includes eleven omnidirectional, megapixel-resolution, color digital video cameras embedded in the ceilings of each room of the house—kitchen, dining room, living room, playroom, entrance, exercise room, three bedrooms, hallway, and bathroom. Fourteen ceiling-mounted microphones are placed for optimal coverage of CD-quality speech in all rooms. When there is no competing noise source, even whispered speech is clearly captured. A server sits in the basement storing all the recordings. They call it the Speechome.

The primary purpose was to get a complete record of their son's language development—every cry, every coo, every "ga-ga-ga" and "da-da-da" ever uttered by the baby while at home, as well as every bit of language input to which the child is exposed. After three years they've collected 230,000 hours of raw data—a truly massive corpus. Compared to this corpus, previous studies are fragmentary at best. Who knows what key moments, previously unknown or overlooked, may be uncovered? The comprehensive nature of the Speechome record will enable observations that were completely impossible to make in the past. For the scientists studying language development, the Speechome approach expands their universe, just as the telescope has done for astronomers.

The size of the Speechome corpus will prove itself typical, if not indeed small, for future research projects. Total Recall will change how scientists learn. And each individual will come to have her own corpus of lifelong learning. Learning will change for all of us.

Technology is already changing what we take the time to learn. We no longer master the slide rule or even trust ourselves to evaluate a complex formula. Instead we turn to our calculators and spreadsheets, both of which we can get on cell phones these days. If you have children, you have probably heard the charge that spelling is an obsolete skill; who writes anything that matters without a spell checker? Even if those who vigorously defend spelling have to concede that it has become a less important skill—after all, as spelling and grammar checkers improve, the product of the poor speller has grown less and less distinct from that of the proficient speller.

Most of us are well along the path of outsourcing our brains to some form of e-memory. I no longer bother to learn telephone numbers; my cell phone remembers them for me. It knows eighteen hundred numbers, far more than I would ever hope to commit to memory, but quantity isn't the issue; I can't even be bothered to memorize the six numbers at my two homes and office. Likewise, there are many facts that need not be on the tip of the tongue as long as they are at your fingertip via your smartphone. The circumference of Earth, the speed of light, the year that Lincoln was assassinated, and Gauss's electric flux law; each of these I once memorized in school. Now each takes about five seconds to look up on my smartphone. None seem quite so important to commit to memory anymore.

This is not to say you shouldn't memorize facts or the correct spelling of words—it's just that such memorization will never again be as important as it once was. And with Total Recall, the list of what is less important to memorize expands to cover everything you know.

Another way technology has already impacted learning is by changing the way we research things. When I was a student in the 1950s and had a paper to write, I'd walk to the library, hunt through the card catalogs and special abstract books, hike around the stacks grabbing items, and eventually sit in the library and make some notes in my notebook for the paper. The next generation had the luxury of photocopying articles, but following up a reference from a photocopied paper required another trip back to the library. All that hassle dampens one's enthusiasm for extensive research.

Nowadays, I can look up anything I want to learn about in an instant. I visit the Web to find those extra references. The Internet allows me to drill down deeper than ever into any given subject. I don't visit our library at Microsoft, and I never ask them for books. I just count on them to subscribe to the online collections that give me access to the professional journals and conference proceedings for my field. They also give me online access to market research, press clippings, and so forth. University Web pages are bursting with all kinds of intellectual treasure. Gone are the days of writing to other libraries to get material not found in my local library. I now call up the thesis of an obscure graduate student from the other side of the world as easily as I acquire a paper by a colleague down the hall. Out-of-print and out-of-copyright books are available by the thousand on sites like Project Gutenberg and Google Book Search. More and more books come straight to my Kindle e-book reader.

With the Web, our ability to research has been greatly amplified. Total Recall will take research productivity another notch higher. Your e-memories will provide quick access to the things you've already seen and the details of what you already know. You can step back from the vast world of information on the Web and focus on what you have found interesting in the past. You can collect and organize your own unique library.

As an individual matures and takes more responsibility for his or her learning, the benefits of Total Recall will multiply. Total Recall will change how we teach students. It will change how we do science, and how all forms of research and scholarship will be pursued. It will change how we learn, from the

simple lessons we absorb in grade school to the wisdom we will
distill in old age.

E-MEMORY IN EDUCATION

Education has been a topic of intense debate since Plato. Should it
be broad and "liberal" or should it be focused on a technical skill
to prepare you for a job? Should education make you a better citizen? Some institutions denounce memorizing and prefer problem-
solving and team interaction. Some promote a classic education,
learning Latin and reading an extensive list of old books. Some
advertise their excellent lecturers, while others say lectures are
passé and point to their hands-on labs. Small class size, notebook
computers for all students, home school, exchange programs, internships, self-directed study . . . the list of approaches is endless.
And while all kinds of techniques are being tried, all kinds of technology will be also applied. Textbooks will become e-textbooks,
most lectures will become e-lectures, and many study groups will
be online.

Nevertheless, disparate as the approaches may be, Total Recall
will have a conspicuous impact on all of them because it provides
an e-memory vessel to hold the knowledge content for lifetime
reference, including all the standard documents of our educational
systems such as articles, books, and class notes.

It will be a new world for the teacher, looking out at a classroom full of lifelogging students. Expectations will change for
students who have e-memories of classes. The e-memories of the
teachers themselves will also impact the way we educate.

Think of the impact on lectures alone. There will no longer be
a question of whether a lecture was remembered; the e-memory of
it will be available at any time. The student will be able to replay a

particularly tricky explanation several times, and to pause at each step to struggle with comprehension; it means the lecture can happen at the student's pace.

In fact, e-lectures are so compelling that students may well prefer them to live lectures. First, they support self-pacing. Additionally, a good lecturer is usually chosen for an e-lecture—you wouldn't bother making it with a bad one—while the quality of instruction out in the world is hit-and-miss. Jim Gemmell has home-schooled his children on occasion, and purchased a set of recorded math lectures that showed an animated chalkboard as you heard the instructor's voice. His children quickly took advantage of the self-pacing. When they later entered public school, they complained about the quality of the teachers, often claiming that they survived entirely on the strength of the previous year's e-lectures.

So, while students may start out recording their teachers' lectures, I suspect that the trend will be more and more toward viewing lectures of the truly great speakers rather than of whoever happens to be assigned to the subject in your local school. Furthermore, I expect the role of the lecture to gradually diminish. An MIT study compared students who prepared using Web-based materials and then heard a lecture with those who prepared using the same Web materials and who then applied a small portion of the material in small projects together with the faculty. The replacement of lectures with hands-on experience led to a 10.8 percent grade improvement.

Of course, whatever live interaction replaces lectures will also be recorded. Students in, say, a physics lab will record video of their pendulum experiment, take a picture of a diagram drawn by the teaching assistant, add some typed and spoken notes, and record the class discussion. The notes they take will also serve as time markers, allowing them to quickly jump to a desired point in the recording. For example, a student will select her note that says

"Weight #2" to begin reviewing the part where she changed the weight on the pendulum.

Imagine you have just returned from a seventh-grade field trip. You had to take pictures of six different kinds of leaves/needles from trees and identify them. The teacher was giving out hints and information as the class walked through the forest. You recorded what you thought you were seeing as you took the pictures. This leaf is prickly. That one is smooth. This one smells like turpentine. You compared notes with your friends and figured out that a couple were wrong, so you added new audio recordings to the pictures, correctly identifying the trees. Back in class the teacher shows pictures the students have taken and the class has to guess the species. Audio is played back to indicate the correct answer. This memorable field trip sticks with you. Years later, when you are forty years old and find yourself telling your daughter about the scaly quality of a cedar, the sounds and sights from that old field trip are what come into your mind—and you bring them up to show your daughter.

I have often been struck by the amount of time students spend comparing notes—not about the class material, but about exactly what the assignments are. I know of one class where assignments were sometimes posted on a class Web site, other times were handed out on paper, and on occasion were even modified verbally. I would call this poor communication and management, but the teacher actually believed it was important for students to struggle through this inconsistency. I don't suppose we will ever be entirely freed of such teachers (or their counterparts in businesses, religious communities, sports teams, and so on). So e-memories will be a big help in tracking what the plan is. Students will be replaying the end of the class to get the new due date for their term paper.

Today, a student may already be creating an education portfolio, that is, a collection of her work chosen to represent her interests and accomplishments. A portfolio may include essays, reports, presentations, videos, or any other sample of the student's work. Hailed for aiding student self-awareness and motivation, in addition to tracking student development, portfolios have been receiving a lot of attention from educators, and e-portfolio offerings have multiplied.

Total Recall will uncover the nuances of each individual's learning style. Some students are visual learners; they don't really get it until they see a drawing. Others are auditory learners who need to hear things. When you record your every move, you can look at what you are doing to better understand what is effective and what is not. Does cramming all night improve your grade, or would you be better off with a good night's sleep? How much exercise, rest, and background music are best for you? What habits make you a more successful learner?

A complete e-memory will allow a much more detailed analysis of a student. Every single answer in assignments and tests can be considered looking for strengths, weaknesses, and trends. The exact time editing an essay down to the history of changes made to each individual sentence can be known.

Teachers will be able to know each student's learning style, and to quickly look up the details of areas that need help. The kind of in-depth insight into a student that might have taken significant one-on-one time to discover will be made automatic, allowing the teacher to go straight to remedial tutoring or to the introduction of an advanced topic. If e-lectures are embraced, and e-memories automatically evaluate the status of students, we could see a radical change in how teachers apportion their time. Instead of spending it dealing with the entire class, teaching to a common denominator and unable to address the needs of both the brightest and the slowest, teachers will be able to concentrate most of their time one-on-

one or in dealing with small groups of students who are roughly at the same level. The instruction will be much more targeted to individual needs.

By pooling data about students' learning styles, study habits, and the results they obtain, we can advance our knowledge about learning itself. Anonymous learning data can be shared to give us a detailed picture over thousands, even millions, of students. The number of students studied and the level of detail examined will be unprecedented. A poor chapter in a standard textbook won't last long, nor will a teaching method that falls down in a certain subject or with a certain kind of student.

Teachers will acquire as much self-knowledge about their teaching as students will about their learning. It will be easier for teachers to repeat their successful moments and to compare notes with one another about the most effective techniques. Suppose you are teaching grade-six science, and you notice that your students did poorly on a certain topic in Chapter 3. You can ask to view a colleague explaining the concepts to her students to see if there is a better way. Or, you may review the relevant assignments and find she has used some supplementary material rather than the textbook. You may recall that a certain unit of history went really well last year, and look back in your e-memories to refresh your memory of just how you taught it.

E-TEXTBOOKS + E-MEMORY = STUDENT MEMEX

Even more certain than the move to e-lectures is the transition to e-textbooks. I remember my Microsoft colleague Chuck Thacker holding up his tablet PC in one of our conference rooms a few years ago.

"Now, if I drop it like this . . ." said Chuck—and then let the tablet crash to the floor.

"You can see I've dropped it on that edge a number of times and the case is cracking."

Chuck was clearly enjoying his job that day. How often do you get paid to drop your PC? As he explained to the group around the conference table, the PC was still running just fine, but he wasn't pleased that the case was cracking. He started explaining the difference in g-forces involved in a landing on carpet versus concrete, and how a little bit of shock absorption protected the normally delicate hard drive. Chuck is interested in dropping tablet PCs because he wants them in the hands of children, and of course, everyone's second question (after the price) will be: What about when they drop them?

Textbooks should be replaced by tablets, notebooks, PDAs, or even cell phones. Rather than lugging a backpack full of textbooks— sometimes so heavy that children have had their backs damaged, leading to some regulations to limit the load—a single notebook computer could hold numerous texts. An e-textbook can be superior in a number of ways. When a student wants to look up a certain passage from his history textbook, he can search for some keywords, rather than flipping through pages. What's the definition of an acute angle? He can find it in a moment. It can include videos of historical events and animations of scientific processes. When he is learning a language, the e-textbook can allow the student to hear the words spoken, in addition to his reading them. Likewise, the computer can listen to the student speaking and comment on his pronunciation. An e-textbook will come preloaded with links. For example, whenever the student is working on a math problem, there will always be a link back to the section explaining the technique. If that section relies on previous material, there will be a further link back to it. E-textbooks will help visual learners develop crib sheets, while auditory learners can create voice prompts for their studying.

While e-books in general have struggled to assure authors that their copyrights will not be violated by pirating, there is no fear that schools will pirate textbooks. We can expect improvement in screens to make reading a better experience, and improved cases that can withstand more abuse, but present technology is already workable.

The switch to e-textbooks is important to Total Recall for two key reasons. First, it means the student will have a recording device with her. E-textbook devices should be capable of note taking, highlighting, picture taking, and audio/video recording. Many, like the tablet, will support handwriting, sketching, and diagram drawing. They will be able to record the student doing her assignments and will capture all the details of how she works. They will know exactly what parts of the text the student has looked at and for how long. Second, the same device that holds the e-textbooks will allow the student to consult her educational e-memories; they will replay class discussions, retrieve notes, or jump to the last point she was reading at.

The ability to consult your learning e-memories is critically different from being able to access textbook or reference material electronically. The vastness of our electronic resources is a wonderful thing, but it is a barrier to refinding, that is, finding something you found before. You may have performed a number of searches and followed several hyperlinks to get to a Web page the first time. An attempt to find the same reference again can be difficult: If you start with slightly different keywords, if you misremember what the top of the page looked like, if you encounter something that looks similar; any of these can prevent you from refinding the page.

In contrast, the pages you have looked at form a much smaller pool to search in. The list of pages that you spent more than ten seconds looking at is even shorter (I have found that to be a very useful filter in MyLifeBits searches—it effectively culls all of the "no, that's not what I'm after" pages). Likewise, a memory of e-textbook use pays off. Few courses cover an entire textbook; sections and even entire chapters are commonly left out of a course. Searching only what you have read before or quickly calling up just highlighted passages makes you much more efficient.

The tablet PC in the hands of the student of the future will be more than just a container of e-textbooks; really, it will be Vannevar Bush's memex. Bush intended memex for scientists, but students need memex just as much. They are collecting material, making notes, needing to look things up quickly, and wanting links to the context quotes are taken from. A student memex is a combination of e-textbooks and e-memory.

A student's memex will be accessible from his tablet PC and their cell phone; it will be with him in class, and everywhere he goes. Classes, lectures, and labs are recorded. When he studies with his friends and is grappling with just how to factor a certain kind of equation, he can bring up the recorded class lecture on his tablet PC and watch the teacher explain at the board again. He is able to quickly review Algebra I in the summer before proceeding to Algebra II in the fall. He can copy and paste from his text and lecture notes to make a crib sheet to study from. He can add his own links, within the text or to other texts. His notes will contain links back into the text. He can highlight and scribble notes over his e-textbooks—and then will be able to quickly find pages based on the amount that has been written over them. Imagine a view of thumbnail pages, only showing those with markup, with the text slightly faded to make the markup more visible. Now the student can quickly scan the pages for familiar markings, which are often more memorable than the actual context of the text—unique doodles and color patterns are encouraged as memory aids. He can also sort the pages being studied by the time spent on each, allowing him to refind a key passage as well as indentifying sections that have been neglected.

Textbooks commonly contain exercises and problem sets, of course, and e-textbooks will too. But the difference is: E-textbooks will be able to record your answers to your e-memory and check if they are right. Adding the e-memory is crucial. Think of a ninth-grade math student. Her assignments should combine older material with the latest lesson, to ensure retention of earlier skills. But which older material? Conventional textbooks make a best guess. An e-textbook with an e-memory knows what older material needs more work. It knows how long since the student last did such a problem. It can drill enough to distinguish between sloppy mistakes and true struggles. Areas of difficulty can be given more emphasis. It may be noted that while completing a certain type of math problem correctly, the student takes a long time, and the student may be referred to techniques that may help, or quizzed on more foundational skills that are suspected to be lacking. Topics that were mastered can be set aside, and only brought back after enough time has elapsed for retention to be a concern. Each assignment is completely customized to the student based on her learning record.

Like the scientists sharing their findings, students can benefit from sharing also. We all recall poorly worded questions or description in textbooks from our schooldays, and needing to pool our knowledge with other students to decipher what was meant. Students will share notes and crib sheets. Group projects will be able to build up a collaborative collection of links and notes from individual e-memories that all can tap into to produce the final result.

Additionally, a student's memex will easily integrate his task list, tracking what assignments are completed and which are due next. It can help manage the student's study habits, for example, pointing out that Johnny spent two hours studying for a history exam worth just 5 percent of the final grade while only spending a half hour on the geography midterm, which is worth 20 percent of the final grade.

HIGHER LEARNING

It is illuminating to contrast the scientist's memex envisioned by Vannevar Bush with what is realized by the World Wide Web. Bush expected that

> Wholly new forms of encyclopedias will appear, ready made with a mesh of associative trails running through them, ready to be dropped into the memex and there amplified.
> . . . There is a new profession of trail blazers, those who find delight in the task of establishing useful trails through the enormous mass of the common record.

Bush didn't foresee the Internet, so he never would have anticipated hyperlinked encyclopedias, such as Wikipedia, Questia, or Encarta, that are used without copying their entire contents into one's personal e-memory. Trailblazers do exist, but in the form of those who publish pages of useful links. I know trailblazing is important, because when my team at Microsoft Research was working on networking, our "trailblazer" page of links to relevant research received far more traffic than any of our other pages for a time.

On the whole, the Web seems more convenient than what Bush imagined, but we do face broken links as pages move, and we can't create links to specific passages on a page unless an HTML "bookmark" is already defined there by the author. Furthermore, copyright concerns come into play if I want to show someone else's page with my markup added. These technical and copyright issues have, so far, prevented the widespread sharing of pages marked up with highlighting and marginal notes. By creating one's own copy, one can prevent broken links and add bookmarks as desired—and share a private copy without announcing a copyright violation to the world. Thus Bush will probably be proved right in predicting that a scientist "sets a reproducer in action, photographs the whole trail out, and passes it to his friend for insertion in his own memex," or, in modern parlance, he e-mails a copy to a friend. (I'll leave it to others to argue the legality of such copies; I merely predict it will happen.)

According to Vannevar Bush, "the inheritance from the master becomes, not only his additions to the world's record, but for his disciples the entire scaffolding by which they were erected." What he failed to see is just how elaborate this scaffolding might be.

To date, it is common for a published paper with a few tables and charts to be the only long-term survivor of a research project that once had volumes of data, "metadata" that describes how the data was gathered, copious notes, and conversations among the researchers. Vannevar Bush saw that more notes and background material might be shared. Jim Gray led the charge in proposing that *everything* could be shared. Think of the amazing detail and enormous volume of data that Deb Roy is collecting. His Speechome corpus need not be reduced to a few publications; the whole data set can be passed on.

Science began with a paradigm of observation and experimentation. Later came a paradigm of theory, and, more recently, a paradigm of computer simulation. The fourth paradigm of science, or the *Gray paradigm,* as I believe it should be called, is a paradigm of data-intensive science. Gray and his colleagues elaborate:

> Traditionally scientists have had good excuses for not saving and documenting everything forever, it was uneconomic or infeasible. So, we have followed the style set by Tycho Brahe and Galileo—maintain careful notebooks and make them available; but, the source data is either not recorded at all, or is discarded after it is reduced. In some cases it is even considered private, especially when done in corporate laboratories!
>
> It is now feasible, even economical, to store everything from most experiments. If you can afford to store some digital information for a year, you can afford to buy a digital cemetery plot that will store it forever. In the future, some fields will no doubt require public storage and access of experimental data. Astronomy is an example of a community that is in transition to this new kind of science and may no doubt be at the forefront because it is traditionally collaborative and minimally funded. Sharing observations is critical and the norm.

Researchers from all fields, not just science, will be able to preserve and share all of their material and notes to the benefit of others. There can be enormous value in a marginal entry indicating that a historical assertion is refuted elsewhere, or a note that the thermometer was slightly moved in 1978, accounting for increased temperature readings, or an explanation of why a certain approach was abandoned. Someone may want to apply a fresh approach to the old data. Shared systems will allow many researchers to pool their material together, so that for some given data, say, an economic report for 2002, you can see comments by many individuals, links to related reports, and metadata describing how the report collected its data and tabulated its results.

Historians ought to jump on the fourth paradigm, and insist on original source material being made readily available. Too many works have relied on secondary sources in the past. And the scope of original sources is about to explode as lifelogging increases. We shall have to see how society evolves to deal with the legacy of e-memories, but I presume that eventually many lifelogs will be opened to a trusted historian to excerpt, if not entirely released to the public.

Suppose someone were to release even a quarter of their lifelog posthumously: It would still confront the historians with a corpus vastly larger than they have ever experienced before. As more people lifelog, historians will also have to delve into the e-memories of other related figures as part of their study.

Earlier, I pointed out that it was a fallacy to worry about having enough time to watch your whole life. An individual would never want to watch his whole life, and knows what he may want to look for in his e-memories. But for the historian it truly is a challenge, because a historian doesn't know what to search for or what can safely be ignored, having not lived the life in question. Thus, historians will become more and more adept at using data mining and pattern recognition, and will come to demand the latest in tools for comparing videos, performing handwriting recognition, converting speech to text, classifying background noise, and much more. They will rely on computing power to help summarize, classify, and identify anomalies, so that they can safely pass over their subject's typical commute to work but not miss the one where she made an unusual stop. Many hours of the subject's life may be classified as "reading," during which time the title of what she reads should usually suffice.

However good automatic analysis may ever be, the most potent historical figures will be, as they are now, studied by many historians. Some will specialize in different periods or different aspects of the figure's life. The subject's e-memories, like the scientific data discussed above, will be in a common repository and the historians' notes and links will be available to one another. Thus, they will not rely just on machine intelligence, but will look to other humans to point out events of significance, to classify material, and to identify trends.

This marked-up repository will also form a new way of delivering history to casual consumers. While only a few experts or motivated amateurs will want to delve into the full data set behind a scientific paper, the appeal of history—of historical stories—is very broad. Many people would be interested in seeing a little more context behind some point of interest, perhaps watching an entire baseball game that includes their favorite major league player as a twelve-year-old, or observing how a politician acted in a press conference early in his career, or listening to the recording of a famous concert mentioned in a musician's biography.

I believe the electronic history exhibits will become more and more faithful to Bush's trails. History presentations will stitch together media into a narrative, truly creating a trail from one artifact to the next, with attached comments. For instance, imagine you are following a trail that I created about the history of computer design. You might start with a chart showing the evolution of computers in the twentieth century, and hearing my voice explain them. Next in the trail is a page describing a computer architecture called SNAP, and you hear me explain Jim Gray's contribution.

But here you grow curious—who was Jim Gray? How did Gordon Bell become involved with him? You see that this page is part of another trail, authored by a fellow named Tom Barclay, called "The Life of Jim Gray," and divert into that trail to learn more about him. Presentations such as these bring Bush's trails together with the World Wide Web to form trail webs, and ultimately a World Wide Web of trails.

Trail webs will have such advantages that traditional museums and science centers will have a hard time competing—what they will do is follow the trend already begun of complementing the museum with computer kiosks, handheld units, and other devices so that the real-space experience can be supplemented by the e-experience. The real-space experience will offer that "wow, I am standing in front of the real thing" feeling, but only the electronic one will let you virtually take apart the priceless artifact and inspect the inside, or see simulations of how it was created. If you have a tour guide at a physical museum you will get only their viewpoint—their trail—through the artifacts. With trail webs, you can listen to the inventor or artist who created the artifact. Many notable experts, each with his or her own unique background and point of view, can take you on a trail through the artifacts.

I have created several e-tours of the Computer Museum's artifacts that are my own interpretation of computing history. Others use the Computer Museum's artifacts as background for their own interpreted e-tours. If you want to see a great demonstration of a web of trails today, try out the World Wide Telescope, where you can take tours of a starry sky, zooming and panning to different areas, reading and hearing commentary at each stop, and, once again, discovering trail intersections that may entice you into entertaining side paths.

LIFELONG LEARNING

Writing this book while using MyLifeBits gives a foretaste of the future of learning. Jim Gemmell recalls reading a paper from MIT comparing lectures with experience, and searches for "MIT lecture web." Too many results come back, so he adds the word "students." Still too many, but now he remembers it was nicely formatted, and narrows down to just PDFs—there it is. He retrieves the paper and rereads the abstract to get his facts straight. In just a couple of minutes the information is included, complete with a full citation and the detail that grades improve by 10.8 percent. His learning about education is leveraged to understand Total Recall.

I imagine a high school senior who has a final exam in biology coming up. To prepare, she looks back to her midterm exams and finds all the problems that she got wrong. She searches her e-memory based on the text in the questions and finds the associated sections in the textbook, as well as the lectures that covered the material. Most were just things she needed to memorize better, so she adds those textbook sections to her "review for final" collection. But there is one problem that she just doesn't understand, so she listens to the critical part of the lecture again, several times in fact, and consults the section in the text as well as looking up material on the Internet. She cuts and pastes key facts to memorize out of her "review for final" collection to create a fill-in-the-blanks study sheet, which she uses to cram for the final.

A graduate student in history has a paper to write on the French Revolution. Her grandfather is a professor of history, and when he did his Ph.D., it was a lot of work to pull together even ten citations for a midterm paper. For the student, every paper and book she has ever read on the French Revolution is instantly available in her e-memory. Besides the main text and four papers she is using in class, she pulls up another twenty-three references that she has encountered over the years. She is able to refresh her memory on a few points and use several quotes from the old papers. What would have taken her grandfather a full day at the library takes her just an hour. Not only is her paper stronger, but her memory of the subject has been reinforced, and her big-picture understanding is broader.

But it is not just students who will supercharge their learning using Total Recall. It is a given today that we all need to be lifelong learners. In fact, many educators insist that just cramming a student's head full of knowledge is not the point; that the goal should be to educate the students to be better learners, brainstormers, and better collaborators. I know that in my own field of computer science, unless one is constantly learning, one's knowledge will quickly become obsolete. Learning may well be the key to the greatest economic rewards of this new technological era.

Imagine Dan, a blueberry farmer. The summer has been very wet, and he consults his e-memories to recall how he dealt with a similar wet season eight years ago. His farm has its own e-memory, which Dan data-mines to understand which varieties of blueberry have been most profitable. Dan is interested in organic farming and has built up his own e-library of Internet articles, recordings of farm visits by a government consultant, and a few talks at the local university. He loves to sit down with his e-memories and a cup of tea to contemplate how he might make his farm better.

Ken is a volunteer hockey coach. He has several e-books on coaching, Internet articles, and notes and recordings from a number of coaching clinics he has attended. He often watches other teams practice and picks the brains of other coaches to discover new drills and approaches. Ken's view of the game borrows from many sources but is unique to him. He loves to mull over his notes and diagrams, to look at video clip examples of certain plays, and then present his insights to his players in a multimedia "chalk talk." Ken loves learning the game and will never stop comparing approaches, connecting dots, and gathering examples.

I could go on multiplying such examples: the mother of a dyslexic child, ever learning about how to better prepare him for the world, the layperson with a deep religious interest, and the countless jobs that deal with rapidly changing information, from lawyers tracking evolving case law to contractors dealing with building codes. Lifelong learning is both a necessity and a joy.

What I find fascinating is that once our learning is self-directed, without any education system telling us what to do, we all act like little professors, just like the scientist that Bush had in mind. We gather material, arrange it, create links, add notes, and generally make sense of it all. We call up bits of it to help put together our next idea.

Arranging one's material is very important. A Total Recall system ought to let you organize, classify, or taxonomize the material you are taking in. I believe there is a very strong case to be made that you don't really have a grasp of your material until you have built a mental model, a structure, such as a taxonomy or mindmap, under which you classify the information being absorbed. A good e-memory will help you arrange material this way for retrieval by classification, and will help you visualize your classification and modify it as your understanding evolves.

A good e-memory lets you step back from information hunting, out in the wilds of the Web, to do some understanding farming back on your home turf. If the Web gets us through the research phase more quickly, if e-memories help us refind and organize our knowledge more quickly, then what do we have more time for? Reflection. Anyone wanting to learn and understand will be doing more pondering, more reflecting, more searching for clues and connections to understanding. Vannevar Bush's imagined memex scientist "ponders over his notes in the evening." So will farmer Dan, coach Ken, and countless others.

Just as the World Wide Web enabled an era of increased research, Total Recall will enable an era of increased reflection.

CHAPTER 7

EVERYDAY LIFE— AND AFTERLIFE

I have a hand-loomed blanket from the late 1800s, handed down from my grandmother Bell. Grandma had seven boys and three girls who, in turn, produced twenty-five grandchildren, including me. The only remaining mementos of her are half a dozen photos, a Bible, a dresser, and the blanket. As the older generation passed away, the Bible was passed on to the remaining eldest son, and eventually on to the grandsons. At each transition, there was inevitably a heated dispute over who ought to inherit the Bible, although none of the clan was especially religious. When it was my turn for the Bible, I declined, but I happily accepted the virtually unknown blanket, which was surreptitiously given to me by a favorite aunt who had been caring for Grandma.

This beautiful blanket, now used as a wall hanging, is one of the few physical artifacts that I care for, but I will soon pass it on to my son. For me, a high-resolution photo will suffice. If I'm feeling sentimental, I'll print a life-sized copy. How can I be satisfied with a digital photo? Wouldn't I take far more pleasure in the actual, tangible blanket? Nothing is quite like the blanket itself, but I have lots of mementos and memories, and taken as a whole, I've discovered that I derive more pleasure from them in digital form. While I am enjoying my e-memories, most people's physical mementos gather dust in an attic—if they even have them.

A typical story of the distribution of physical items after a dear one's passing goes like this: "I spent a day being careful to preserve and bundle my mother's artifacts (mostly correspondence and scrapbooks) for my sister. After a day my filter became very narrow and all the stuff just went to the dump. And my sister never came to claim what I had saved."

Physical family heirlooms and mementos pass down through random branches of the family, eventually arriving where they are unrecognized, unappreciated, and discarded. Other relatives, who might have been keenly interested in the mementos, are unlikely to even know of their existence. At best, only one person has custody of the valued heirloom, which requires physical space and careful preservation if it is to be passed on to the next generation.

How refreshing to contrast this to a digital legacy! All of the heirs can have a copy, it can be quite expansive, and in the event you don't especially want to know anything about the departed, there is no cost to your space or attention.

Of course, a family heirloom may be some rare antique or a diamond bracelet. Objects with monetary value will be squabbled over until the end of time. But I am speaking here of emotional value. Not long ago, Jim Gemmell overheard a woman speaking to her friend:

> I'm going to give you my cell phone number . . . you can't leave messages on my home phone because I've saved too many messages from my grandson.

This lady wasn't just keeping track of when her grandson had called or maintaining an accurate record of the words he said. For her, hearing his little voice was precious.

This is a very different aspect of Total Recall from what we have discussed in the last three chapters. This kind of memory gets to the heart of your emotional life, to the fabric and climax of the best stories you can tell about yourself. Sure, we can have perfect recall with regard to our work, our health, our pursuit of knowledge; we will improve our minds, bodies, and ventures. But this also touches our hearts, our emotional lives, with all that impractical stuff that makes up the rest of our days and nights. It is an awesome prospect that even these kinds of memories can become e-memories, totally searchable, even ready for scientific analysis. What would Proust have made of it?

These kinds of memories so often exist between you and someone you are close to. Between you and your grandmother, son, friend. They are often family memories. As such, the bio-memories overlap. Each individual's enhances the others'. In the world of Total Recall, your e-memories will supercharge this enhancement. Your personal relationships inside and around your family will be transformed.

If we can have a complete record of the things about people that especially provoke meaning for us, what will we do with this complete record when they are gone? We will maintain the e-memory of that person as a treasured heirloom. And, someday, we will ask it questions. The e-memory will answer. You will have virtual immortality.

RECORDING AND STORYTELLING

Lifelogging implies collecting countless digital artifacts—and increasing the variety of artifacts is all to the good. We want all the strands of the fabric of our life. Vacation videos. Snowy dioramas from that skiing trip. Our first blanket (or our grandmother's). Songs we wrote in high school. Birthday cards. Tickets to concerts. What your father said in that crucial moment of the third quarter in that crucial game. Maps of our travels. Recipes. Laundry lists. Lists of guests invited to a party. Toasts, eulogies, and your baby's first attempts at speech. The fabric of your personal, intimate life.

These e-memories are so enjoyable when they come up on screensavers that I declared screensavers the "killer application" for e-memories (ahead of search, which so many people presume is the chief end of Total Recall). The MyLifeBits screensaver shows photos and also selects random ten-second video clips out of longer videos. Before Total Recall, years would have passed between those occasions when I bothered pulling out old videotapes to play selections from them; now I can enjoy clips from them daily.

Screensavers are pleasurable on my desktop PC and notebook, but they really shine on my big plasma screen. I am tracking the progress of organic light-emitting polymer technology and looking forward to the day that an entire wall can be a screen. I want to sit at my dining room table and have my wall transport me to some other place or cherished past event—perhaps I can feel like I am reliving a train ride through the Rockies or sailing on the San Francisco Bay. I also love little "picture frame" screens that can sit on an end table and, instead of being stuck on one photo as real picture frames are, root through my e-memories to show all kinds of things.

Having different types of artifacts contributes to the richness of your e-memories. Think of the lady I mentioned above with her phone messages from her grandson. Jim Gemmell has some treasured audio recordings of his grandfather's Scottish brogue. Video is essential for "preserving people," a view that I have only recently come to appreciate. Seeing someone move, speak, and make facial expressions provides a distinctive view of a personality that written records and photos can rarely touch. Now add location, temperature, heart rate, and other new values we will be sensing, and really fascinating perspectives are formed.

With location tracking, we can plot you in space as well as time—and that's exactly what we did with the MyLifeBits Trip Replay program. Gemmell came back to work one Monday, after spending the weekend at a sports tournament with his son, traveling with a GPS and taking pictures with his digital camera. When he joined us in the office, we were able see on a map everywhere he had been, and where his photos had been taken. Trip Replay even animated his travels, showing him moving around on a map, with photos flipping up as they occurred.

Gemmell was excited. "Look at this!" he exclaimed. "And think about it: I have the information about each game my son played in my calendar. All the components are here to tell the story of my weekend without me doing any work at all!" We started brainstorming about how this could be automatically wrapped up in an attractive form and sent to his parents to fill them in on their grandson's weekend. We later hired an intern to prototype a system that let you just select a time range, exclude a few duds or embarrassing moments, and then click "Blog it."

Consequently, I'm bullish on automatic travelogues. I envision a service that takes your itinerary and produces a trip log with photos of where you stayed, sights that you passed by, meals eaten, et cetera. With the use of GPS, the entire trip can be created in great detail that might even exceed the actual trip experience. As I write, Telestial is preparing to roll out a service that will track where your cell phone is, and post stock photos of the locations you visit to a Web site you designate. You can add captions by sending SMS text messages from your phone. This is just an early (and very simple) entry in a coming wave of automatic travelogue offerings.

The landscape of our e-memories becomes lush as we share with one another. The value of pooling media is already evident in photo- and video-sharing Web sites, like Flickr and YouTube. Facebook shows us how much we enjoy having others' comments on our photos. Think of an extended family gathering for Grandma's birthday with a number of people taking photos and maybe shooting a little video. Once the media is shared, if there is even one keener in the family who adds comments, flags key moments, or groups the media in some way, then we all benefit.

Imagine the parents of players on a basketball team. Suppose the video of the game is posted to a Web site, keyed with the score clock, allowing everyone to quickly jump to a certain point in the game and skip breaks or dull parts. After the game, each family goes online and flags a few favorite plays made by their child. When all their annotations are added together, an automatic highlight reel is easily produced. Years later, a player can go back to the copy of the game in his e-memory and relive his big moments. I believe that many sports venues in the future will install cameras and post video automatically. They'll make some extra money, and as we share our thoughts on the sport with one another, some splendid e-memories will be constructed.

Furthermore, sharing is necessary to fill in a key missing part of any e-memory. If I wear a camera capturing my own point of view, there is always one person left out of the footage—me! You

and I must share our point-of-view footage with each other in order
to appear in our own lifelogs.

One thing we will certainly share with each other is stories.
Humans are storytellers, and no matter how much I value a recording, I'll always love to hear someone else tell the story of the
event.

In 1989, when my ninety-year-old mother visited me, I asked
her to write some stories to pass on to her four grandchildren. In
particular, I asked her to tell stories of the changes she had wit-
nessed since her birth in 1899. She wrote about social life, clubs,
church, and school. She told stories of Christmas, Thanksgiving,
farming, gardening, and food from butchering to canning. Just recently, my sister thanked me for initiating these stories, as she was
reading them to her own grandchildren.

Jim Gemmell was thinking about stories in 2005 when he asked
me to spend a number of lunch breaks with the half-dozen others in the lab answering questions about my experience at Digital
Equipment Corporation. We ended up with an hour of video stories and opinions never expressed elsewhere. I'm sure the viewer
would get a different sense of me from seeing these videos than he
would by reading what I have written about life at that time.

I believe oral histories are irreplaceable. While I love all the
artifacts at the Computer History Museum, I've come to see the
stories they collect in oral histories and public talks as the most
important thing the museum does. Of course, memories fade
over time, so for accuracy we want more than just oral histories.
Still, inaccuracies and all, I want to capture people recounting
stories.

For Total Recall to fulfill its full potential, people must be able
to tell stories anytime, anyplace, any way they feel like it. This
could be by sitting in front of video camera each night for a dear-

diary entry, talking about some episode while driving, or typing in some thoughts about a recorded event.

AUTOMATIC SUMMARIZATION

With today's technology, your e-memories would be a mixed blessing for your heirs. They would have the benefit of more knowledge about you, but it would come to them as an enormous, daunting mess. Your heirs may enjoy looking at random photos, or searching for e-mails containing the names of presidents in order to read some of your political perspectives, but they will likely miss the most important and interesting bits, and may be too intimidated to spend much time with your e-memories.

I felt this way about my own early scanned collection, and it was my frustrated eruption of "It's just bits!" that galvanized Jim Gemmell into action and got us started on MyLifeBits. Thankfully, our prototyping work has assured me that things will get much better.

We will see the evolution of software that will reduce the chores involved in making one's life bits worthwhile to others. It will help to develop tools to make storytelling and human arrangement easier. But fully automatic approaches are even more important. Some of this we get quite readily just by storing more information together. For example, your e-mails, calendar entries, Web-page visits, and digital photos all have time stamps that readily lend themselves to time line displays.

Time lines are a really compelling way to visualize your life, and software can help automatically produce digestible time lines. Our colleague Eric Horvitz and his research team have done some very promising work in predicting which events people will consider to be significant "memory landmarks." This allows the best material to be put on a time line, and the less interesting material

hidden away until it is asked for. Eric demonstrated the software
to a reporter, starting with pictures of his wife and son:

> "What's cool—I love this feature—I can say, 'Go to July
> Fourth,' and it's making guesses about the things I am likely
> to remember, to use memory landmarks, and it jumped right
> to this place," he said. The screen showed several images—a
> small-town parade, and his wife and son among figures at a
> cookout, from July 4, 2005. Responding to his request, "the
> computer brought up its best guess."
>
> "It comes to understand your mind, how you organize
> your memories, by what you choose. It learns to become like
> you, to help you be a better you."

Remember the Dublin City University project that finds novel
SenseCam images out of thousands to highlight the interesting
ones and spare you from the mundane? Automatic summarization has become an entire field of research. There are even sub-
specialties, for example, summarizing just video. While we would
like it best if a human creates a photo album, it is possible for the
computer to do a pretty good job of automatic photo-album com-
position, choosing only interesting, high-quality photos. The nice
thing about machine-composed albums or time lines is that, unlike
physical albums, they keep the "outtakes"; with just a few mouse
clicks you can retrieve the other shots that weren't included in the
album, allowing you to be as absorbed as you like with one particular event or topic.

With automatic summarization, posterity will be able to
browse your e-memories starting from a manageable "birds-eye"
view of a life, rather than just confronting an intimidating jungle
of material.

IMMORTALIZING JIM GRAY

When Jim Gray went missing in 2007, I was not alone in wishing to immortalize him in the most rich and resilient way possible.

I am certain that Jim Gray's name will be immortal at least in some ways. His name cannot be neglected in any history of computing as a winner of the Turing Award (often called the Nobel Prize of computer science). He is best known for his role in developing transaction processing, which we all use every time we withdraw cash from banking ATMs. In an effort to have his name be even better remembered, I helped establish the Jim Gray Endowed Chair in Computer Systems at UC Berkeley. I'd also like to see a building named after him. Jim's astronomer friends have already identified an asteroid that will bear his name.

For a computer scientist like Jim, the most common way to gain an immortal name is to pass on ideas that are used by future generations. If you are lucky, some concept will be named after you. Moore's Law is undoubtedly the best-known, predicting that transistor density in computer chips would double every two years, and explaining the meteoric rise of computing power. I hope that someday we will refer to the *Gray Data Cube*, the *Gray Transaction Processing Benchmarks*, the *Gray Five-Minute Rule*, and the *Gray Paradigm of Scientific Discovery*.

I'll certainly be lobbying for such an immortal name for Jim. However, while it would be fitting for him to join the ranks of other such esteemed names, having an immortal name is a pretty superficial immortality. We may say Pythagoras' name until the end of time when discussing geometry, but we will never know much about him, or even how he did his work. I want something better for Jim.

Someone's work can be immortalized, as in the paintings of

the great masters, buildings by a brilliant architect, or some notable equation. Going deeper, the *way* they worked may be immortalized: their techniques, their approaches, their professional relationships, and the stories of them at work. For instance, we know a fair bit about the work of Isaac Newton, including the story of him in his early twenties, going to the countryside to avoid an outbreak of the plague and, like any typical young man with too much time on his hands, whiling away his time—inventing calculus and discovering the law of gravitation.

Jim Gray's Web site reveals a lot about him as a computer scientist. The extensive publications on the site reveal his drive for understanding through experiments and measurements. However, the site is missing those additional, critical stories that help us understand the man at work. Other sites, like the National Library of Medicine's "leaders in biomedical research and public health Profiles in Science" Web site (http://profiles.nlm.nih.gov), provide a slightly more personal look. At this site, luminaries such as Francis Crick (who discovered the structure of DNA) have archives that include articles, bibliographies, books, brochures, certificates, drawings, exams, interviews, lectures, letters, notebooks, photos, and schedules. Still, we end up with only a fragmentary view of their lives.

Jim Gray's family, friends, and colleagues are the sole repository of what he was really like: how Jim would slap your back with bubbling enthusiasm when he congratulated you, or how, when he thought your ideas were nuts, he would politely pronounce himself "puzzled" and furrow his brow. His Web site doesn't tell you of his countless lunches on the least expensive sandwiches in San Francisco, despite his wealth. Then there are all the stories from his family, his sailing buddies, friends from college days, and others.

The story of Jim Gray is spread out on the computers he used, in his personal effects, and in the e-memories and bio-memories of those who knew him. Doing justice to his story means bringing them all together and presenting them in a comprehensible way.

MY MEMORIES OF JIM

After losing Jim, I naturally reminisced about my own relationship with him. A quick search in MyLifeBits turned up the following items that reference Jim in some fashion: 13,000 e-mails, 1,600 Web pages, 100 presentations, 289 photos, 600 documents, several videos, and a phone call. I don't recall what first prompted Jim and me to get together, but MyLifeBits has a copy of the penciled calendar entry from 1994 for our first meeting. Other calendar entries include Jim taking me out on his sailboat—the same one he would later vanish in.

In 1994, Jim had just finished four years heading DEC's San Francisco Lab on Market Street and had turned consultant. Since 1989, I had been a Silicon Valley angel investor and a consultant to Microsoft Research and others (since I didn't spend much time consulting, some friends kidded me that *consultant* was a code word for "unemployed"). Our first meeting at my Los Altos home revealed our shared views on the importance of industry standards and an approach to increasing computing power via many cheap PCs working in concert together. We found that we both preferred small teams and esteemed building influential prototypes. It was the beginning of a stimulating collaboration and a heartfelt friendship.

After being an independent consultant for a while, Jim felt that he needed the confines of an organization, and he convinced me that I needed more structure too. He had been talking to Microsoft. We believed Microsoft was the place to be because of how we felt about standards and leverage, and moreover the respect and enjoyment of the community we would be part of. I jumped the gun and e-mailed the Redmond folks to hurry up and start a Microsoft Lab in San Francisco for Jim:

Sun Jan 08 15:41:55 1995
To: Rick Rashid; Nathan Myhrvold; . . . Dave Cutler

From: Gordon Bell
Subject: Approaches to Servers and Scalability . . . and an AD Lab here!

>Folks, Here's how Jim Gray and I see the next decade or two: A Scalable Network and Platforms (SNAP) architecture is predicated on one set of standards: an ubiquitous ATM network <u>and PC-sized platforms</u>. SNAP allows upsizing i.e. building world-scale computers from a single platform architecture in a scalable fashion. SNAP will encourage further industry de-stratification. It eliminates the traditional computer price class distinctions (mainframes, minis, PCs) and goes a long way to eliminate the stratified business models of traditional computer suppliers. SNAP will cause a computer industry upheaval greater than the early 1990s client-server downsizing wave. That wave created a large UNIX market displacing IBM mainframes and proprietary minis. But the UNIX market is fragmented and small when compared to Compaq and NT. UNIX would have to consolidate around one or two dialects in order to get the volumes required to compete with NT. This seems improbable, so Microsoft's NT is likely to become the dominant server standard for all hardware platforms, just as Windows garnered the desktop or client side.

Jim sent his own e-mail, pointing out that he had not "put me up" to writing mine, and enumerating the difficulties of operating a remote lab. However, he strongly validated the outlined vision. Microsoft liked the idea, and Jim's Bay Area Research Center (BARC) opened in the summer of 1995 in San Francisco. I was honored and delighted to join the lab in August of that year. We hired Jim Gemmell to work with me that fall.

Though the BARC lab peaked at only around ten members, it had an impact beyond its numbers. Aerial imagery of the world was brought to the Internet by the BARC Terra Server, which led to the Microsoft Live Maps site and predated Google maps by five years. Later, Jim would turn the view up to the heavens, and work on the Sky Server project. His broad agenda got him involved in such far-flung projects as the "land speed record" for network transmission and fail-safe databases. Meanwhile, Gemmell and I were working on telepresence: putting a conference on the Web, playing with altering someone's gaze direction in video, and shipping new network protocols in Microsoft operating systems. Later, of course, we got into MyLifeBits.

A memorable event was in May 1997, when Jim gave an on-stage demo with Bill Gates, using more than a hundred PCs to achieve one billion transactions per day. I also recall Jim's glee on April Fool's Day 2005, when he had just finished measuring a half-billion transactions per day using his relatively old laptop. He wrote a report observing that a common PC could execute eighty times "more than one of the largest U.S. bank's 1970s traffic—it approximates the total U.S. 1970s financial transaction volume. Very modest modern computers can easily solve yesterday's problems." The data and report illustrate Gray's fondness for understanding through constant building and experimentation.

Through various paths, Jim infected me with the importance of data. It's "all about the data," he would say. In one of our more playful times, while discussing how to get the concern for data into the national computing resource allocation agenda, we bumped into John Markoff, a friend and columnist at *The New York Times* who also had an office in our building. We made the case that the national computing agenda was missing the point by just thinking about computation speed. John took our picture in the lab on Friday and wrote an article that appeared in the *Times* on Sunday. Our compute-centric friends in Washington were not especially happy, but they slowly came around to our view, and after many years the situation is gradually changing from models and simulation to real world, data-based, data-intensive science.

DIGITAL IMMORTALITY

I've just told some stories about Jim Gray, and shared an e-mail artifact. But if my publisher would let me, I'd probably include a photo; and in the e-book, some audio and video too. If Jim had lifelogged, there would be a wealth of e-memories to peruse.

A complete lifelog is an awesome prospect. Passing on your e-memories might be seen as a sort of one-way immortality, allowing you to communicate to the future. But Jim Gray and I used to speculate about the possibility of something really worth calling digital immortality, where a digital version of yourself lives on and interacts with posterity. What if, a hundred years from now, your heirs could ask you questions and you could answer?

You might be familiar with the word *avatar,* meaning an image or computer-generated character that represents someone in cyberspace. An avatar might be a picture of you that appears while you instant message with someone. Or, it may be a three-dimensional character in a virtual-reality program like Second Life. Someday, those who lifelog will be able to create avatars that do an amazing job of impersonating them.

Imagine if you could have a conversation with an avatar of your great-grandfather, seeing his face, hearing his voice, and having your questions answered in your great-grandfather's unique turn of phrase. It seems clear that at least some aspects of your person can be expressed as a program that actually interacts with future generations. Lifelogging is the key.

Your lifelog will have all the details of how you sound, the phrases you employ, questions you have answered, and facts about your life. It will contain recordings of you under stress and relaxed, pleased and annoyed, in triumph and in defeat. It will know your favorite quips and mottos. It will have pictures of you that can be overlaid on a three-dimensional model. If someone should want to impersonate you, your lifelog will be the ultimate gold mine.

There has already been some progress making avatars that answer as the individual would have. Carnegie Mellon University researchers have created a program that lets you interview Albert Einstein. I've invested in a company called MyCyberTwin, whose software trains an instant messenger (IM) avatar how to answer as you would. You just give it content from your own IM conversations, e-mail, blog, et cetera, and it learns how to imitate you—it becomes your "cyber twin."

MyCyberTwin analyzed transcripts of the television show *The Simpsons* to create a cyber twin for Bart Simpson, who had the following chat with me:

You: What's up?
Bart: Well, I was interested in getting the fire hydrants from the school to power my downhill racer—wanna help me?
You: No. How is Lisa?
Bart: She can be so annoying, but she is still pretty cool. Do you like my Mom's hair?
You: Blue hair is pretty cool too. What do you think of your dad?
Bart: My dad, Homer . . . he got a chess set once, then packed it

away. When I asked him why, he said he is saving it for a brainy day.

Have you been married to my aunt Selma?

You: No. How old are you?

Bart: I'm ten, man. Best age in the world: You can do what you want without getting in too much trouble.

You: What do you want to do when you grow up?

Bart: It's always important to consider the consequences of anything you do.

Are you a friend of Principal Skinner, by any chance?

You: Do you have a pet?

Bart: Who needs pets when I have Homer? Actually, I do have Santa's Little Helper too. He's a dog. What pets do you have?

MyCyberTwin specializes in creating chat avatars to perform customer service, for example, answering questions for bank customers. Their avatars have gotten better scores than humans in accuracy, sales performance, and customer satisfaction. Now the MyCyberTwin folks are intrigued by the idea of taking my own e-memories as input—there is enough of what I have said in e-mail, letters, chat, papers, and so forth, that one ought to be able to construct a pretty realistic Gordon Bell cyber twin.

Alan Turing, a founding father of computer science, proposed the Turing test for determining a machine's capability to demonstrate intelligence: A human judge has a conversation with a human and a machine, each of which tries to appear human. If the judge can't tell which one is human, then the machine has passed the test. Turing proposed typewritten exchanges; we can update that to computer chat without changing the essence of the test. Thus, we can have a cyber-twin test: You chat with someone and his cyber twin. If you can't tell the difference, then the machine passes the test. Note that the cyber twin could have a much better memory than the human did; it should be taught to forget in a similar way to the human for real simulation. But it should remember when we really want the answer! As I write, there is a fair bit of work remaining before any cyber twin could pass the test, but substantial progress seems likely.

I see four steps in the progression of digital immortality. First is digitizing the legacy media one has. Second is supplementing one's e-memories with new digital sources. The third is two-way immortality—the ability to actually interact with an avatar that responds just like you would. The fourth is an avatar that learns and changes over time just as you *would have*.

Having an avatar that actually learns and grows over time is a much more speculative idea. Who's to say that it is working correctly? If we could predict how someone would behave after death, then we could predict people during life—an idea that sounds farfetched. Perhaps a more realistic goal would be to shoot not for growth—as in change—but just accumulated knowledge, so the machine would recall when it last communicated with you and what was said: "Hi Gordon. I talked to you yesterday. You told me about your vacation."

While I believe the fourth step will remain science fiction, an avatar passing the cyber-twin test is not. A lot of sci-fi and artificial intelligence discussion is about true machine intelligence, where programs actually learn and grow. Personally, the more I learn about machine intelligence, the more I am impressed with the learning ability of the average two-year-old. True machine intelligence remains elusive. Computers can beat human chess masters now, but the computer's greatest advantage is its ability to enumerate each and every possible move and outcome; most of us would call that tediously mindless, not intelligent. Similarly, in an attempt to answer factual questions like "What year was Abraham Lincoln assassinated?" brute-force approaches that just scan many encyclopedias and newspaper sources looking for common words often do as well as or better than programs that attempt to parse and comprehend the same texts. I am confident about avatars passing the cyber-twin test because I see that lifelogs will contain enough information to support such brute-force approaches. No sci-fi, truly intelligent machines are required.

By carefully mining your lifelog, we should be able to ask your e-memories questions and hear your answers. We can change the game from a search to a discussion.

Humans have a natural propensity for recording life. Just look at all the people walking around with cameras and video cams. You'd be hard-pressed to find a home without photo albums, home movies, scrapbooks, and mementos. The one thing many people would be sure to rescue from the flames of a burning home would be their photo albums. We love to reminisce, and if you think of all the photos and home movies taken, it seems we enjoy *enhanced reminiscence*: not just remembering but also hearing and seeing recordings or artifacts from the past. A few of us go beyond just confining ourselves to recordings and objects, and actually edit movies, or create scrapbooks with captions and artistic layout. Some even take classes from companies like Creative Memories to learn to do it better. The rest of us envy them the time and talent to produce such compelling stories.

Your e-memories will prepare you for the digital afterlife. Already, for a fee, Web sites like www.legacy.com and www.for evernetwork.com offer to store letters, essays, photos, videos, and stories to pass on to future generations. Famento.com helps people share family stories, building e-memorials to loved ones. These sites are the digital equivalents of cemeteries and libraries.

Imagine opening up an old tomb and finding it full of historic artifacts. That would be interesting for a while, but just think of how much more interesting the artifacts would be if they came with their own curator, ready to help guide you through and explain them. That's what I expect Total Recall software to do for our e-memories, with automatic travelogues, automatic summarization, and cyber twins. Today, as I struggle with setting up Jim Gray's digital legacy, I know we've got a long way to go. But I also know that Jim would have insisted on seeing this as a challenge—another chance to do and apply science by creating understanding and something of value.

PART THREE

PART THREE

CHAPTER 8

LIVING THROUGH THE REVOLUTION

New technologies have always forced mankind to adapt to new realities, from iron tools to mobile phones. The changes worked on our societies by powered machinery were so radical that we refer to the industrial "revolution." We are now beginning the Total Recall revolution.

I'm a technologist, not a Luddite, so I'll leave abstract discussions about whether we should turn back the clock to others. Total Recall is inevitable regardless of such discussions. However, as a realist I also know that we must come to grips with the implications of our technology. Some implications I see as "bugs" to be fixed. For example, an unresolved bug of the industrial revolution is pollution. Other implications are simply changes that we must adapt to, such as modern transportation implying that one can commute to work and get fresh fruit from another continent.

This chapter is about the changes generated by Total Recall, both the bugs to be fixed, and the adaptations that will be required. These changes are mostly about what happens to our e-memories once we have them. Could we lose them? Could they fall into the wrong hands? What is their proper use? Different cultures may come up with different answers. Technology will yield unintended negative consequences and pleasant surprises too. If the answers aren't all clear, most of the questions are.

DATA LOSS AND DECAY

Right now we face data decay and loss. Data often only exists in one place, so a crash of the host device means permanent loss of the data. Files formats may become unreadable over time. We need improved data longevity.

One morning in the fall of 2008 my notebook computer wouldn't start up. I was in Australia and it was all I had. I broke out in a sweat. Questions fired off in my mind: Is the hard drive shot? How much isn't backed up? How will I get my e-memory back in action? I decided the biggest problems would be a presentation and some articles if the disk was bad, because they had no backup. I also faced the hassle of paying bills without Microsoft Money, which I've come to count on.

Fortunately I had backed everything up onto my assistant's machine in San Francisco two weeks before getting on the plane. I could download my e-memory as of two weeks ago from her, or I could just use it straight off her machine when I was on the Internet. My most recent e-mail was intact because it was on our corporate e-mail server.

This episode reminded me that I need to be prepared to lose whatever is not backed up. I had just finished authoring a presentation and was crazy not to have copied it to a USB thumb drive.

I believe the chief obstacle to data longevity is low expectations. For too long, too many of us have been content to see the data in our PCs, PDAs, and cell phones as transient. We shrug at losing the phone numbers from our cell phones, or not being able to do anything with files on an old floppy disk. This tolerance for data decay was natural when personal computers were new and few people had any experience with electronic storage. Fortunately, a couple of decades into widespread computer usage, we are learning better. I'm encouraged by signs of this trend, for example, seeing that Dell notebooks ship with Dell DataSafe Online software installed to perform backups to their site.

Protecting against outright data loss involves two techniques: replication and backup. *Replication* means that a copy is made of every bit of data you own. The more copies, the better. It is best to make copies that are located far away from each other, so that a hurricane, earthquake, or fire doesn't destroy all of the copies at once. Such geographic replication has been commonly employed by Fortune 500 companies for many years; a bank cannot tolerate even the thought of losing all its account balances.

Backup is a little different than replication. A replica is important, but what if you accidentally change an important file? The next day you look at the file and realize you've wiped out some valuable information. You can't turn to a replica, because it has faithfully copied your destructive changes. A backup is a snapshot of your data at a given moment, to cover you in the event that you need to get back to an older version.

Consumer software to perform replication and backup is readily available and even free in some cases. Pretty good solutions are already in place. Higher demand will give us all the solutions we could want at an affordable price. The bug of outright data loss has already been fixed with replication and backup, so we just need to ask for it—and the more of us who ask for it the better and cheaper it will get.

However, outright loss is not the only threat to our data's longevity. We may also experience data decay. Suppose you are user of SuperPhotoEdit version 3.0 and you create a collage of family vacation pictures. Ten years later, you launch version 8.3, and try to load the old collage, only to see "File format not supported." Or, even worse, you have a new computer, and have no desire to buy SuperPhotoEdit. All you want is to see your collage, but you are two hundred dollars and a half hour of installing away from that.

Will your data be readable fifty years from now? Jim Gemmell and I posted some audio files on the Web in 1997 and about five years later they couldn't be played. The team at Microsoft in charge of such things explained that their license for the format had expired and the company that had the rights to the format had gone bankrupt. It was illegal to make the clips playable, with no real likelihood that the company would ever be resurrected to make it legal again. It was a dead format.

I call this the "Dear Appy" problem, after a flight of fancy in which I imagined poor forlorn data, utterly abandoned, writing a letter to the application that created it:

Dear Appy,

I thought we had a commitment. You were going to understand and support me forever. What happened? Where are you?

Signed,
Lost and Forgotten Data

A really complete solution to Dear Appy would be able to emulate any hardware, operating system, and application for all time. Then you could run the old program and open your file. That isn't going to happen, but most of what we want and need is not rocket science; it is possible with a little care and, again, by our demanding new software and services rather than being content with the status quo. I'll cover some practical steps for today in Chapter 9, and look into the future for Dear Appy in Chapter 10.

DATA ENTANGLEMENT

No one can take away your bio-memories, but some of your e-memories might not even belong to you. Were I to resign from Microsoft right now, they would immediately demand that I perform a partial e-lobotomy, removing all work-related e-memories.

When Jim Gray went missing, there was a fair bit of consternation about what to do with his notebook computer. It was loaded with all kinds of Microsoft information, some being proprietary to the company. It also contained quite a few photos and a fair bit of correspondence that was quite personal. Microsoft wasn't so sure it would be a good idea to give Jim's wife, Donna, access to the information. What if she saw something confidential? Donna felt uncomfortable having Microsoft employees looking at it before she did. What if they saw something very private? Microsoft, having possession of the machine, had the advantage. It took nearly a year for Donna to be given the data that Microsoft had deemed as being fit for her eyes. Donna naturally wonders if some things were deleted (perhaps by mistake) that need not have been, and if some things were seen that she wished to remain unseen. If Jim had left it at home, the situation would have played out in the same way but with the roles reversed.

It will be interesting to see how society adapts to e-memories in the workplace. Surely there will be an evolution of law and employment contracts—which, in day-to-day practice, people may pay as little attention to as they do to posted speed limits. How many of the millions who will legally commit to delete e-memories will actually do so, given no possible way for anyone to ever verify if they really did it? Contracts may stipulate nonretention of e-memories, but any teeth in such agreements will be regarding disclosure, not retention.

I can't imagine maintaining separate computers for work and personal memories. Even having separate e-mail accounts for all the different organizations that might want me to purge certain memories would be ridiculous. I have a separate personal e-mail account, but I receive personal e-mails on my work account and work-related e-mails on my personal account. My calendar is an intermixing of work and personal life.

My data is entangled.

I try to organize everything I have to separate work memories from personal ones, but it's tough. I know I will end up with information on the work side that really is part of my personal story— for example, hotel and airline arrangements for my business travel. Likewise, I no doubt have recorded chat sessions with Jim Gemmell that include a few lines about, say, company reorganization in the middle of stories of our daily lives.

If I care most about leaving my story to posterity, I'll err on the side of marking things personal. If I care most about not ending up in a lawsuit, I'll err on the side of calling items work-related. I can't think of how to make things better, apart from improved tools for marking items as work or personal. Maybe that's a bug we can fix. Maybe that's just reality we must adapt to.

ADAPTING TO MORE SELF-KNOWLEDGE

One change we will have to adapt to is having vastly more knowledge about ourselves. I've already covered how this self-knowledge will improve such things as health. But some people have shared with me a worry that they may learn things about themselves that they don't really want to know—the depressing truth may get out. They go further than the Soviets, who erased what they didn't like from their history; these folk would erase everything *just in case* there might be something they don't like.

They ask: Do we really want to know all this stuff? Liam Bannon, writing in favor of forgetting, offers up the inarguable: "More data do not imply better-quality decisions." Of course that's true—but flawed human memories do not imply quality decisions either.

There are many instances where you need more data to get a better picture of things. One example would be tracking your heartbeat for an entire month so as to not miss a few key events. For people reviewing performance or progress, an accurate record can make all the difference over a fuzzy and rationalized memory.

In the world of business, we do not hear arguments against record keeping or concerns that facing the truth is inferior to a comfortably dimmed memory. Accountants do not spend their lunch breaks debating the need to forget or whether storing every single transaction might clutter the record too much. To the contrary, it has become established business practice to write down clear and measurable goals, to measure your performance, and then to look back at your performance compared to your predictions to see how you did.

In sports, athletes carefully record their batting average, save percentage, race time, or whatever measure applies to them. They don't rely on their memory of how they tend to play toward the end of games; their fourth-quarter statistics are compared to other quarters. Even in youth sports, elaborate statistics are kept and young players desiring to improve their game watch videos of themselves with commentary from a coach or trainer.

The question may well be: How much truth can you take? An athlete may feel uncomfortable watching a video of herself using incorrect technique, and the salesman may squirm to look back on his projections, but such is the price of self-improvement. Successful people don't shy away from the honest record. Management guru Peter Drucker relates this to a person's career, saying:

The only way to discover your strengths is through feedback analysis. Whenever you make a key decision or take a key action, write down what you expect will happen. Nine or twelve months later, compare the actual results with your expectations. I have been practicing this method for fifteen to twenty years now, and every time I do it, I am surprised. The feedback analysis showed me, for instance—and to my great surprise—that I have an intuitive understanding of technical people, whether they are engineers or accountants or market researchers. It also showed me that I don't really resonate with generalists.

Feedback analysis is by no means new. It was invented sometime in the fourteenth century by an otherwise totally obscure German theologian and picked up quite independently, some 150 years later, by John Calvin and Ignatius Loyola, each of whom incorporated it into the practice of his followers. In fact, the steadfast focus on performance and results that this habit produces explains why the institutions these two men founded, the Calvinist church and the Jesuit order, came to dominate Europe within thirty years.

Imagine being confronted with the actual amount of time you spend with your daughter rather than your rosy accounting of it. Or having your eyes opened to how truly abrasive you were in a conversation. Right now, only very special friends could confront me with such facts in a way I would accept. And they receive my thanks for helping me grow as a person. In fact, for such a mirror of ourselves, we sometimes pay such special friends and call them therapists or counselors.

It's up to you: You can tackle as much or as little truth about yourself as you have the stomach for. In court, we ask for the truth, the *whole* truth, and nothing but the truth. It might be painful, but I believe better memory really is better.

ADAPTING TO BEING RECORDED

Of course, having Total Recall to help with your self-awareness is one thing; having a spouse drag up e-memories to berate you is another. Even worse, imagine a moment of weakness being posted to YouTube by a bitter former friend. The Total Recall revolution implies that others are recording just as much as you are. That's a big change to adapt to.

The world is already adapting to being recorded. Google has cars drive down streets with a 360-degree camera on the roof of a car to create their street views for their maps. As soon as they were launched, street views prompted an outcry by people concerned that they would be shown in places or situations that were embarrassing to them. Sure enough, street views have included such things as men entering strip clubs, the view up a girl's skirt, a man relieving himself against a bus, and a police bust. Canada's privacy minister warned Google that street views may be illegal there, and many other countries have raised legal questions. The U.S. military prohibited pictures of military bases. In response to these reactions, Google began blurring faces and license plates in the pictures, and also has taken down many of the embarrassing ones (however, copies live on elsewhere in the Internet). The world is still adapting to street views.

Wearable computing pioneer Steve Mann has a unique response to being recorded: He "shoots back." Ever since he was an MIT graduate student in the 1980s, Steve has been wearing a computer-and-eyewear combination that captures the light that would have gone to his eye, sends the signal to a computer, and then presents a computer-processed image for him to actually see. His view can thus be altered or replaced entirely.

While Steve's apparatus is much more than just a recording device, it can record, and this has gotten him into trouble. Once he was forcibly ejected by security guards from the Art Gallery of Ontario, under the rationale that he might infringe the copyright of the artwork in the museum. This and similar incidents have, ironically, increased the amount of recording Steve has done. Now, in order to capture any incidents of violence, he always records in such situations.

Steve takes particular exception to being told not to record in places that have surveillance cameras placed on him. "It seemed that the very people who pointed cameras at citizens were the ones who were most afraid of new inventions and technologies of citizen cameras." Fair is fair—can't he shoot back? He even coined the termed *sousveillance,* from the French to "watch from below," in contrast with *surveillance,* French for "to watch from above."

Steve turns the tables on the surveillance folks in many ways.

If they are concerned about him violating copyright of their art, he wears a T-shirt with artwork on the front and requests they turn off their cameras to avoid infringing his copyright. He is often told surveillance is for his own safety, and replies that he is recording for safety too—would they be willing to sign a form taking responsibility for the consequences of removing his "safety device"?

These days, no one can tell if Steve is wearing a camera, anyhow; he appears to be wearing ordinary glasses and his computing equipment is of the pocket-PC variety. Without any suspicions raised, he is free to shoot back to his heart's content. Big Brother, meet Little Brother.

Soon, of course, will come the multiplication of Little Brothers, recording all over the place. And where there are e-memories, e-gossip can't be far behind. E-gossip is progressing from text like "I saw Gordon do X" to the actual e-memories in pictures, audio, and video. For all the talk of Big Brother, Little Brother is more likely to impact you.

There are many implications to believing what you do may be recorded—and replayed. It could put you on your best behavior. Antisocial behavior could be exposed and condemned. You couldn't expect to get away with many lies. There is even some cold comfort in knowing that if I use my e-memories to harangue you over something you've done, you will have a copy of my harassment to use against me. Crimes could be caught on tape—while I was writing this book, patrons of Oakland's subway used their cell phones to record video of a man being shot by a police officer.

On the other hand, not all secrets are nefarious; I may be sneaking out to buy you a gift. People may be inhibited from going for needed treatment if they think it may lead to exposure of their problems. And relationships could become stilted, with candid conversations being replaced by the excruciatingly careful speech we are used to hearing from politicians, who are the first wave of society to have their words regularly recorded and played back to them.

So how will we adapt to being recorded? Will it be a free-for-all? Will we pass more restrictive laws? Recording someone talking without their permission is already illegal in many places. Customs will no doubt evolve regarding when it is socially acceptable to record. In one culture, it may be good manners just to let people know you are recording. Another culture may deem lifelogging in the company of others an absolute taboo. Birthday parties might be fair game, while first dates are not.

I think requiring consent to record will be the likely direction of custom and law, and technology will be developed to this end. When several of us gather, our devices will communicate information about who is allowed to record whom. If I chat with Ted and Mary, Ted may consent to my recording while Mary does not. My log of the conversation would then have all images of Mary blurred and all of her speech erased. At the same time, she might have allowed Ted to record her.

New sorts of relationships will arise from the adoption of lifelogging. There will be those you trust to record and not divulge. There will be those you trust to not record. Perhaps making promises on the record will become a milestone in relationships.

I am not dogmatic about absolutely continuous recording. I think we will sometimes shut off the recording and say things off the record. We may even occasionally stop recording just to have the "novelty" of memories that exist only in our heads. Even so, if people only logged a tenth of their lives, the changes in society would still be dramatic. And even a tenth of a life logged would be enormous and significant—how I wish I had a tenth of my grandfather's life. Once we get a taste of a tenth we will want much more.

ADAPTING IN COURT

Could your e-memories be forced to testify against you? Richard Nixon tried the route of plausible deniability, saying, "You can say 'I can't recall . . . ,'" but tape recordings of his conversations demolished his denials.

Today in the USA, you can be compelled to produce a diary as part of discovery in a court case. If e-memories are considered a digital diary, they would surely be treated the same way. However, a recent court case ruled against the state's being able to compel a man to divulge an encryption key for his hard drive, explaining that it would violate his right against self-incrimination. As the case wound through the courts, one judge opined, "Electronic storage devices function as an extension of our own memory. . . . They are capable of storing our thoughts, ranging from the most whimsical to the most profound." This case gives us a glimpse of hope that the law may eventually come to protect our digital memories. Falling a little short of this, some lawyers are arguing that searching one's hard drive should be considered especially invasive, more akin to having your body searched than having your papers searched, and thus requiring a higher standard of justification.

I believe that some sensitive information will be stored in a "Swiss data bank," an actual offshore, encrypted, secret account, which you can plausibly deny the existence of. It may take having several such accounts, so that if evidence was unearthed indicating that you had one, you could turn over the least sensitive. Furthermore, just as secret agents sometimes use a code phrase to indicate they have been compromised, there may be an optional password to the Swiss data bank—intended to be handed over to the authorities—that digitally shreds or somehow hides away some key pieces. It could also function to add all kinds of erroneous data throughout the store, putting the veracity of any of it in doubt.

My advice on hiding information is amoral; it can be used for both good and bad. I'm not aiming to help the next Nixon or pedophile (the court case regarding the encryption key involved a man who had child pornography on his hard drive). Those who commit illegal or immoral acts may be best served by actually deleting their records, but I really don't care if they get good advice or not. I do know that protecting data can be essential to the man holding a Bible study in his house in China, or a homosexual in Iran, both of whom face government persecution. By helping protect such individuals from their tyrannical governments, we can also ensure that liberal governments don't have the chance to become more tyrannical. That's the spirit behind the United States Fifth Amendment, and I want to see the law, society, and technology move in keeping with that spirit.

After hearing one of our lectures on MyLifeBits, it is pretty common for people to express other concerns about having their e-memories used against them. What if the GPS record of my position over time is used to infer that I was speeding? Could I get a ticket? What if the health information I am tracking shows the likelihood of a medical condition—could my insurance company use that as grounds to cancel my coverage? The complete answer to these issues will take time to develop, and will span technology (like the Swiss data bank), law (such as a recognition of the right to not testify against oneself digitally), and business (such as a recognition by health insurance companies that they save money, due to improved overall health, by guaranteeing continued coverage to those who track more information). We will gradually adapt.

ADAPTATION STARTS NOW

Many of the changes coming are not just Total Recall issues, and they're not just out in the future. People often query me about the security of my e-memories, and wonder whether I've created a treasure trove for hackers and identity thieves. They don't seem to realize that their present, tiny, e-memories already contain everything an identity thief needs—and their physical recycling bin is an even more attractive target.

Regardless of law and custom it will become harder and harder to know for sure if you are being recorded, due to continued miniaturization. You *might* be recording me; I can't be sure. And I will act just the same thinking you might record me as I would knowing for sure you will record me. Steve Mann demonstrates this in an experiment called MaybeCam. Steve and friends go out in bulky shirts with a dark Plexiglas panel in the front—similar to the domes that hide surveillance cameras—and printed with:

> For **your** protection, a video record of you and your establishment *may* be **transmitted** and **recorded** at **remote** locations.
>
> ALL CRIMINAL ACTS PROSECUTED!!!

Some of these MaybeCam shirts have cameras behind them, some do not. People react as if they are being recorded even if none of the shirts have cameras.

We will cross the threshold into living much of our lives with the possibility of being recorded long before lifelogging becomes mainstream. And there is already more than enough value in your present e-memory to warrant replication and backup. E-gossip is well under way; the Internet is already teeming with compromising home videos and photos. Were no further progress to take place toward Total Recall, the need to adapt in every area I've discussed is just as strong and urgent.

Life with Total Recall seems as alien to us now as the first automobiles must have felt to the horseback riders they roared past. But Total Recall, like the automobile, is rejected only at the price of giving up great advantages. The greatest failure of adaptation would be the failure to take advantage of the new technology. How sad it would be to lose memories or to fail to pass on a digital legacy. What a mistake to miss out on improved health, learning, productivity, or self-awareness. We need to adapt to reap the benefits.

CHAPTER 9

GETTING STARTED

A revolution is coming, sure enough, and all over the world people are gradually drifting toward it, recording and storing more and more of their lives. But what if you want to push ahead and experience more of the revolution sooner? You won't have a research team to take you as far as I have gone, but you can still do quite a lot. You can begin your own Total Recall revolution right now. What follows in this chapter is a plan of action for getting the revolution started in your life, and maybe beyond.

This chapter is about getting started with what we have, even if we have to employ the occasional "hack." I'll mention a few products, but the technology is moving so quickly that I suggest paying attention to what these devices fundamentally accomplish, not their bells and whistles or brand names. And of course, check in on totalrecallbook.com for the latest updates on services and products that we think mark progress in the march to Total Recall.

STEP 1: THE SETUP

Approach your e-memory with a plan and you will get better results. Decide which aspects of your life you want to be able to recall, from physical objects like bowling balls to born-digital items like e-mails and GPS locations.

As I described in Part One, when I started my own Total Recall project back in 1998, I had boxes of papers in storage, foldersful of paper in file drawers, and piles of papers all over my home and office. My shelves overflowed with books, LPs, CDs, VHS videotapes, and DVDs. My photographs were archived in shoe boxes alongside slides in projector carousels. Memorabilia was displayed in curio cabinets and my refrigerator displayed my grandchildren's latest masterpieces. Just looking at all my stuff was a bit overwhelming, but I decided to go at everything.

You may wish to embark on your e-memory project in a more focused way. But the overall system of recording, storing, and using will remain essentially the same. You may consider, as a friend of mine did, putting together a multigenerational list of all the recipes your family made and/or recorded. My friend found some as far back as five generations. She scanned and/or retyped the recipes—if you go back five generations you can expect a thousand or more—and she included any comments from the recipe's creator. So there you have a specific set of data recorded and digitally stored. Now. How do you access it? How do you use it? How about a cookbook of recipes with eggs? Or given my health concerns, one without eggs? In fact, that friend made a cookbook that was such a success with her family members that she has begun a second Total Recall project to create a family Web site that will contain the cookbook, all family photos, handwritten letters, blogs, full genealogies, birth and death records, health problems, and so forth.

No doubt for many people, family is a forum in which Total Recall technologies meet their warmest welcome.

When my son-in-law, Bob, embarked on his Total Recall project, he decided to focus only on photographs. After his old pictures are all scanned, he will be tackling his music. His approach is to break down the entire project into manageable sections by data type: first photos, then music, then videos, then paper, and so forth.

Maybe you want to limit your e-memory at first only to family, or food, or music. Maybe you want to limit it to your health, or your work, or your romantic life.

THE BASIC EQUIPMENT

If you don't already have the tools of this sort of digital enterprise, you will need to buy them. Most are now fairly familiar in American households. If you think this is just buying more clutter, keep in mind that these tools will easily fit in the space you make in your life by reducing the paper and memorabilia that surrounds you. Moreover, the Total Recall revolution is being built on the strength of a few key fundamental devices and they are all fairly small.

A smartphone

Your cell phone should be a smart one, that is, one that also performs the functions of a personal digital assistant or PDA. At a minimum, it should support: phone, text messaging, instant messaging, camera, Web browsing, e-mail, reminders, and synchronization of your contacts and calendar from wherever in the cloud you keep them. If you can, get one that supports GPS. Music playback is nice too.

Many will let you connect to local Wi-Fi network hotspots when you are in range, giving you a faster connection (and sometimes avoiding airtime fees). The major platforms (Symbian, iPhone, BlackBerry, Windows Mobile, and so on) have all kinds of applications that you can install to help you with things like: diaries, health records, lists, time cards, financial records, customer relationship management, expense tracking, and banking.

The smartphone is an absolutely critical tool for lifelogging because you tend to carry it everywhere all the time. It takes pictures, records audio comments, and allows you to type notes and to-do items. It tracks time and place.

A GPS unit

A GPS built into your smartphone might be all you need—as long as you have the software to retrieve the GPS data out of your phone. Many phones ship without such software and it can sometimes be hard to find, even if it exists for your phone. If you can't get GPS data from your phone, then buy a GPS unit even if you have it built into your camera. After all, you want to know where you have been even when you weren't taking pictures. If your camera does not have GPS built in, then a separate GPS unit is essential for supplementing your pictures with location information.

I've used some Garmin GPS units, and also tried out the Trackstick. Personally, I hate changing batteries, so the rechargeable i-Blue 747 is my current favorite. Whatever GPS you get, make sure you can save files in a standard format like GPX. Check that you can actually read the text in your file using a Web browser or a text editor like Notepad or Word. That way you will always have latitude/longitude coordinates in a format you can access.

A digital camera

Pictures taken by smartphones are better than no pictures, but at present only dedicated cameras provide really acceptable quality.

The good news is that virtually any digital camera these days takes decent pictures and—very important—stamps the date and time on each photo. At the time of writing, only a few cameras support built-in GPS, but if you can afford one with GPS, get it—because where something happened is just as important to your memory as when it happened. I hope by the time you read this that GPS in cameras is mainstream, and I am willing to bet that by 2013, nearly all cameras will have GPS. If you don't get a camera with GPS, then make sure not to skip the purchase of a separate GPS unit.

A personal computer

Laptop or desktop, your personal computer will be essential. If your computer is more than a couple of years old, consider buying a new one. Not surprisingly, for Total Recall, memory is a critical component. Buy a computer with as much disk memory as you can afford; at least a hundred gigabytes for a laptop or three hundred gigabytes for a desktop computer. Purchase an external disk of five hundred gigabytes or more for backup. Get the latest-generation operating system so you will have integrated features like desktop search and photo-tagging. Apple's Leopard and Microsoft's Vista can do this for you.

An Internet connection

An Internet connection is essential so that you can take advantage of online billing and other items that are born digitally—each one represents paper that you won't have to scan. Paying for a faster link in the upstream direction (from your home to the Internet) may be helpful for backing up to a cloud service, or making your files accessible from a home server. But high bandwidth isn't essential if you won't be moving a lot of data from your home to the cloud.

A scanner

You need a scanner that can digitize anything you have on paper: memos, letters, health records, pictures, slides, business cards, and so forth. Scanners that handle multiple sizes and types of paper are worth the extra expense. Scanners should allow you to digitize one piece of paper or a stack of papers, simultaneously scanning both sides.

I use the Fujitsu ScanSnap desktop scanner. It's nice and small, and I find it so handy that I have one at work and another at home. It lets me stack in pages and scans both sides at once directly into Adobe's PDF format. The current generation of desktop scanners comes close to meeting my ideal that scanning a document be as easy as discarding it. Ultimately scanners will be so reliable that we can confidently shred the document the moment the scan is complete. But we aren't there yet.

A flatbed scanner is great for mementos, like medals, plaques, and so on. Unlike a digital camera, it always gets the lighting right. Some stuff just won't fit in any scanner, though, so sometimes you will have to use a camera; if you can get it outdoors on a cloudy day, you can often get it nicely lit without reflections.

Finally, make sure your scanner software is performing optical character recognition (OCR) on the scanned pages so that later the computer will be able to search for the text inside them.

DEALING WITH WHAT YOU ALREADY HAVE

Properly equipped, you are now ready to convert your old analog life's worth of papers and memorabilia to digital form.

Set a goal of being paperless within a year. Besides scanning the paper you already have, you should also arrange to receive more born-digital communications in the future, to reduce the flow of paper that you need to scan. Request that all statements, invoices, and other communications be delivered online. Your phone company, utility company, cable provider, and other services you do business with should be happy to stop sending you paper (and paying for postage), and if they aren't they'll probably be out of business before long. Many of the communications you receive online will be in PDF format, so just click the save button. Some, however, will come up inside your browser as a regular Web page—I'll explain more about saving Web pages below.

If you have many years' worth of paper accumulated in file drawers, cabinets, and boxes, consider sending it all out to a service for scanning. It will likely cost between four hundred and a thousand dollars for ten thousand pages. The price increases for mixed sizes and artifacts such as scrapbooks. If you have a lot, and you value your time, I'd call it money well spent.

Even if you use a scanning service now, you will still need a scanner for future incoming paper. Some organizations aren't prepared to go paperless, so you will keep getting items to scan from them. Set up a dedicated scanning in-box, and don't let it get more than an inch deep.

Today's scanners and software do not automatically add tags or keywords to your documents, and are not likely to for quite a few years. Make sure that OCR is being performed on every scan. And follow my tips below for file naming and tagging.

Occasionally, I find it handy to use my camera instead of a scanner. For instance, while I'm traveling, I might snap a quick picture of a receipt and tear it up rather than take it home to scan. I also hate keeping around the big cases holding my software CDs—but I can never peel off the label containing the product key without destroying it. My solution is to snap a picture of the label (the lighting only has to be good enough to let me read the product key).

This way, I was able to recycle a bunch of cases and cut back to one compact little holder of CDs for all my software.

Books

For your books that are out of copyright, the chances get better every day that you will be able to download a copy from the Web for free. Project Gutenberg has more than twenty-five thousand free books available in their catalog and more than a hundred thousand available through their partners and affiliates. Google is scanning millions of titles; keep an eye on them to make more books available. Several formats may be available; pick either PDF or some other text-friendly format so you can search the entire book with your e-memory.

Online libraries are available for a fee. Questia, the largest online library as of 2008, has books, journals, magazines, and newspapers all available and searchable via keywords. They even provide a personal library shelf, where you can store books that you've viewed. The workspace allows you to create and manage projects, track your research by viewing highlighted passages, notes, and citations you've made, and even make instant bibliographies or source notes. LexisNexis is another online service that provides business, legal, and news services, all searchable easily though keywords. Whenever possible, save copies of what you read, or at least a note of what you read. For example, if you copy and paste a passage into OneNote, it will also save a link to the Web page.

Invest in Amazon's Kindle, Sony's eBook, or another similar electronic reading device. That way, your new books can be born digital. You also will be able to mark passages and load them into your computer. Wizcon makes a pen scanner that permits you to make highlights in any of your reading materials and automatically transfer those highlights to your computer for future reference.

When you buy a new appliance, go online, download the electronic version of the manual, and throw the paper one away. You may end up having to scan a few manuals, but this happens less and less as every year goes by. Manufacturers, after all, are happy to have you download a file rather than phone them and take time from their staff.

For e-books, e-articles, and e-manuals, the issue is searchability. You want to be able to recall passages that you have read before. With downloaded manuals, you can search on your PC. LexisNexis would require you to return to their site to search if you haven't saved a copy of the article you want. Refinding a passage from a Kindle book could mean going to the Amazon Web site if the book is no longer on your Kindle. Still, with a little effort you can achieve nearly Total Recall with your reading material.

Address books, calendars, and reminders

If your address book has cross-outs five layers deep, or you have a calendar hanging in your kitchen with birthdays, anniversaries, other important dates, and appointments, or your to-do list consists of sticky labels papering your office walls, you need to go digital and it's not as hard as you think.

Start with your address book. You can choose from a host of already available applications. Eudora, Outlook, Outlook Express, and various freeware programs have contact, calendaring, and time management systems that connect to corporate mail servers or public mail services like AOL, Google, or Hotmail. Apple's OS X has integrated calendar and address book applications. Many of these applications can synchronize with the information in your cell phone or PDA, so that device can remain updated too.

If you are using a cloud-based system for your contacts, calendar, or reminders, I strongly recommend using a client such as Outlook so that your e-memory has a copy of everything that's in the cloud. This will protect you from a service that may lose your stuff (perhaps by going out of business). It also gives you a copy when your Internet connection goes down.

Many of your contacts are born digital. For example, you receive a call on your cell phone with a number you and your cell phone don't recognize. Once you take the call, your cell phone asks if you want to store the number as a new contact. This is easily done and when you sync your phone to your computer, your computer now has the new number in your address book.

IBM has a program called Pensieve (named after a stone receptacle for storing memories, à la Harry Potter) to manage business contacts. After you use your cell phone to snap the photo of a person you meet along with his or her business card, you enter the information into your computer. The program syncs that data with the date, time, and information in your calendar for when you met that person. When you search for someone, you enter one bit of data and up comes photo, name, phone number, fax, company info, and so on.

Nokia is taking this idea one step further, allowing their cell phones that have GPS and a compass to become full memory aids by using images taken from the cell phone. Anything you see, a person, place, or thing, is snapped as a picture and tagged with location. This new phone will be preloaded with tags for places and things in a set of cities, allowing travelers to easily become accustomed to their new environments.

I can't say enough about the importance of the calendar to mark life's minor and major time-posts that are likely to be useful for recall. Use your calendar not just to schedule upcoming events, but also as a diary, putting entries in even after the event so that your calendar is a complete record. Every birthday, celebration, dinner, and meeting should be noted.

Pictures

If you have lots of pictures, slides, or negatives, send them to a service. The drugstore Walgreens scanned over two thousand negatives for a friend in less than twenty-four hours. In 2008, Scan-MyPhotos.com would scan a thousand photos for fifty dollars. The more you have, the steeper the discount.

Unless you are a serious photographer, beware the "photo scanner." I've checked out "high-end" photo scanners, hoping they would help me scan faster, only to learn that "high-end" in a photo scanner often means doing fancy smudge and scratch removal and actually requires more of my time to manage the process. The lower-end scanners I've tried have been better at feeding through batches of photos, but unfortunately, they have also been cheap enough to break down.

I've found the best scanners are the ones that are multipurpose and handle paper, photographs, and business cards. If you have a lot of slides, you can find a special slide scanner, or possibly an attachment for your paper scanner to help handle slides.

Sometimes you will want to scan a photo album without taking the photos out. There is no choice but to use a flatbed scanner for that. Don't try using a digital camera unless you have extraordinary skill in photography and lighting.

Music

Music CDs, of course, can be easily converted ("ripped") into your computer. For old formats like LPs and tapes, you can do it yourself to convert to digital format. It's not really that hard, but it can be finicky. You'll have to get the right cables to connect, say, your turntable to your computer's sound card, and you'll have to set the levels carefully to get a good digital recording. There's software out there that that will spot the gaps in the music and break the whole down into individual songs, and will even help with hiss reduction and make labeling easy (I have used the Windows Plus package for this in the past). If that sounds beyond your technical ability or patience, chances are the shop in your neighborhood that digitizes old videos can also digitize your music for you.

Use a music database such as iTunes, Windows Media, Winamp, or Zune to organize your music. The database is part of the player. As you rip new music, these databases will automatically catalog it and create the file-and-folder system for you. It's powerful and allows you to control how you want your music inputted, sorted, and displayed. It will also go out on the Web to find the CD artwork, if available.

Movies and Film

For old films and home movies, I once again recommend using a service, such as iMemories.com. Slidescanning.com does home movies as well as slides. If you are a techie, there are solutions out there and you can mess around with and make them work—the guys on my team have. But for most folks it's just not worth the hassle. Send them to a shop and get back the DVDs.

The real trouble with movies is that we are still a few years out from having enough storage space to rip your whole collection into your computer, as you do with your CDs. For the time being, you are going to be stuck with some DVDs and tapes disturbing the feng shui of your otherwise decluttered life. It is well worth ripping a few of your favorite moments, though.

Virtually all items can be destroyed after digitizing (archivists hate it when I say things like that—but they will have to give up a few papers to get thousands more scanned pages). Of course, some things will be kept even by the most energetic lifeloggers. You may have to keep papers that have intrinsic value, like stock certificates or autographs or, say, original sheet music by a famous composer. You may plan to keep a photo album and enjoy it until it falls apart and fades—in any case, you should rest assured that you have the digital version forever. I recommend that you develop a way of marking digitized items so that you don't mess up and do the work of digitizing them over again. For instance, I mark the pictures that I keep with an *S* on the back, indicating that they have been scanned. You may want to put a small Post-It note on the underside of a trophy that you have a good picture of. But if you ask me, nothing beats the feeling of feeding your paper to the shredder and seeing your clutter evaporate.

STEP II: LIFELOGGING

Now that you've focused on the potential for Total Recall to improve your life, have the tools of the trade, and have begun getting rid of all that paper and other junk, it's time to start recording more of your life digitally. Time to get lifelogging.

I own several digital cameras because of trade-offs in features and size. If I could only have one, I'd pick a pocket-sized camera so that I would be more likely to carry it around with me. If not for my pocket camera, I'd have no shots of my ride in the cab of San Francisco Fire Department Aerial Truck T-13 (what a thrill!). If you don't have your pocket camera with you, take a snapshot using your smartphone. The quality may be lower, but at least you'll have some visual e-memory. The bottom line: Carry a camera and snap away.

I believe in taking video "cliplets"—little clips of five seconds or less. Five seconds is often all it takes to capture the ambience of a moment. No photo, no matter how good, can convey the movement like the five seconds of hula dancers I shot in Hawaii. A quick shot of my grandson saying hello is priceless. And sometimes I like to swing the camera around in a panorama in an attempt to capture the feel of a place I'm in. My camera and cell phone are fine for video cliplets, so I don't bother with a video camera much. With longer videos, I have to be concerned about space on my PC. So most long videos remain in DVD form, but all my video cliplets get added to my e-memory.

Remember when people put pins in a map to show where they had traveled? You can do it digitally by collecting global positioning system (GPS) tracks. I've made trip records of walks, car rides, train rides, and airplane rides. The GPS comes with me into the wilderness and into the skies thirty thousand feet above the Pacific Ocean. You either need a cell phone with GPS records that you can export from it, or a stand-alone GPS unit that you can carry in your pocket.

If batteries are a hassle, and you use your car a lot, it may suffice to have a car charger and just leave the GPS in your car. This at least gives an approximation of where you are, based on where you drove to.

In addition to creating a record of your travels, a GPS is used to add location information to photos. As I'll discuss later, manual labeling of photos can be a lot of work, so having time and location correct on every photo is critical. If you don't have a GPS camera, then you need to ensure that the date and time on your camera are set correctly (actually, you should be sure of that anyhow—you will be really glad to have the right time on all your photos). Now, if you know from the GPS where you were at a certain time, and you know when a picture was taken, you can infer where the picture was taken. The location information can then be inserted into the picture file alongside the date and time. This is called geotagging or geolocating your photos. You can do this with Microsoft Pro Photo Tools, HoudahGeo, or many other programs available on the Web (your GPS device may come with such software).

An alternative to GPS is a memory card with built-in Wi-Fi networking for your camera. This allows you to wirelessly transmit your photos back to your computer or an online photo-sharing site. More important, the card can geolocate every photo—so long as there are Wi-Fi signals at your location.

For audio recording, I carry an Olympus WS-320 audio recorder (though changing batteries is a pain). In a pinch, I also record using the audio recording function in my cell phone. For recording meetings, I like to use OneNote and record directly into my PC. To save telephone conversations, Skype calls can be saved automatically. Recording from a cell phone or home phone is presently more complicated, and there are legal issues; I'd wait for that space to evolve a little more before you jump in.

HEALTH DATA

Health logging is going to rapidly improve in quality convenience in the next few years. Do everything you can to get involved in this potentially life-changing and life-saving trend. You are the only one with access to all of your health data, so take ownership of it and collect all that you can. Take advantage of new technology that helps you achieve quantitative health.

Start out by creating a simple document for medical information about you and any family members you want to keep track of. List all the immunizations, allergies, medications, and any important events, for example, when a surgery was performed. Whenever you get a simple test result—say, a blood pressure value—add it to the document. Aim to have all the key statistics about your health in one quick reference document.

A couple of years ago, I visited the Canyon Ranch wellness center in Lenox, Massachusetts. After a host of blood tests, body scans, cardio stress tests, exercise evaluations, and even gene reports, I finished my stay with more medical and fitness information about myself than I could ever have imagined. The information that was most interesting was the fitness information. I've used this as a foundation to compare with fitness facts I record today.

I wear a Polar heart monitor strap whenever I work out. This allows me to capture heart information and compare it to what I've received at Canyon Ranch and subsequent tests with my cardiologists. My trainer has created a program tailored to help with my heart and with my core. (Supposedly we have balance issues as we age that I clearly observe.) After scanning this into my e-memory, I can now track my progress and it is easy for her to make changes to the training program electronically.

Dr. Christiane Northrup's research has found that walking ten thousand steps a day helps your heart stay healthy. I now wear a pedometer, which downloads what I walk each day into my e-memory. Some days are better than others, but over the last year or so, I have actually averaged about ten thousand steps a day. Dr. Northrup would be as happy as my heart is.

You should determine the area of your health that is most important to you, and get a device that helps you quantify your status. For example, to track your fitness, you might get the BodyBugg arm strap that I mentioned in the health chapter. If you are concerned about your blood pressure, you should buy something like the Omron HEM-790IT blood pressure monitor, which will enable you to upload data to the Microsoft HealthVault.

I've already advised you to get rid of all your paper, so don't get too upset when I tell you to acquire even more paper that you will need to scan. If you really want total control of your health information, first you need to obtain it, which, sadly, nearly always means paper. You will need to contact every doctor, specialist, dentist, hospital, or health facility that has a record of you. You will need to write or fax them to obtain a copy of these records; a phone call won't suffice. Keep a list of all of them and check it off as you receive each record. Also, you will need to do the same to obtain your medical insurance "explanation of benefits" forms, if you don't have them already. Although this aspect of Total Recall took the most time for me to accomplish, having these records and measurements at my fingertips has saved my life after my double bypass redux.

Quicken Health is a database on your own computer that keeps track of the paper blizzard that is typical of a chronic condition or a major procedure. It holds all the letters, bills, and insurance documents—it tracks the money flow and who paid what, when. While such systems are substantially more detailed than financial transactions, they are more than a decade behind the financial industries in terms of their ability to handle health transactions in a humane way. This program was created by a frustrated Quicken employee who saw it as the only way to follow the paper.

With paper under control, you can move on to electronic information. Some physicians communicate via a proprietary e-mail system. You should keep copies of these conversations for your life bits. As more providers in your health-care network go digital, be sure to ask if you can download copies of reports, prescriptions, X-rays, or whatever they will let you have.

TAKE NOTES

Lifelogging will be increasingly automatic. However, right now there is no replacement for just recording a few phrases or sentences. Doing this is more comfortable in a work or educational setting, but whether it is a "note to self," notes about a meeting, or an epigram you want to remember, make a note and give it a date and a place.

In meetings, I like to use OneNote. I can type, or I can use handwriting and draw things using my tablet computer. OneNote can record audio and will sync the audio to the notes I am taking. So, I can later click on a note and have the audio play from the point in time when I made the note. OneNote is also really great for copying passages out of Web pages and keeping a link back to the page you got it from.

For a quick typed note, there are several options. Sending yourself an e-mail works well. You can leave the e-mail in your in-box as a reminder, or put it in a folder of notes on different topics if that was your intent. I use Outlook, which has both notes and tasks; other productivity tools should have similar features. Tasks differ from notes by have properties such as due dates.

For spoken notes, I used to use the record audio note feature on my smartphone. These days, I prefer using reQall. With reQall, I dial a certain phone number, and am prompted, "What would you like to do?" I say, "Add," and then, after a beep, say my note. That audio recording gets delivered to my e-mail as an audio attachment, and a transcription is made of what I said in the text of the e-mail. I find this especially handy when I'm driving: I just hit a speed dial button, talk to reQall, and then I know I will have the note in my in-box to deal with later. I used to forget more ideas than I remembered, but now it's easy to save them. There's more to reQall than I have space to cover here. It understands reminders and shopping lists. You can send it messages from instant messenger, and it can send you reminders in SMS text messages. It really is a fantastic tool for e-memory.

Evernote is another powerful e-memory tool. It covers a lot of ground, from clipping Web pages on the PC to audio recordings on its cell phone software. What I think is particularly interesting about it is how it does OCR when you take snapshots. It does a great job of detecting text in a picture, allowing you to take pictures of stuff from wine labels to whiteboards, and then searching for the text in them.

At a dinner party recently, a guest demonstrated his latest find: a digital pen called a Pulse Pen from Livescribe that both records the lecture and allows you to take notes that can be uploaded to your PC. I know when I sit through lectures, many times the speaker is faster than my pen. The Pulse pen solves this problem because if you miss a word, the pen captures it. The only negative is that it requires special paper.

Other electronic pens are made by IOgear, whose Mobile Digital Scribe pen doesn't require special paper, but also doesn't record the audio. It also requires an extra pager-size device to upload the information to your PC. The ZPen from Dane-Elec is like IOgear's device and doubles as a one-gigabyte flash drive.

TELL STORIES

Advancing from raw media to stories doesn't have to be as time consuming as it is for novelists and filmmakers. We are all storytellers, it is just that we can't all be Shakespeare, or Toni Morrison, or Steven Spielberg.

Begin making dear-diary e-memories, just like you take notes.

Send yourself an e-mail recounting the humorous quip your nephew just made. On your way home from the ball game, call reQall and talk about what happened. Point your camera at yourself and record a short video clip telling about someone you just met. Knowing that you keep an e-copy of everything also means that any stories you send to others become part of your story. So send more e-mails to family and friends telling stories.

Some cameras let you add audio comments to the pictures you take. Take advantage of the feature so that the picture comes with you telling the story behind it. Even if you don't have such a camera, you can always add your voice later. Photo Story is a nice application that lets you create a voice-over story with pictures that zoom and pan, giving you that "history channel" feel. VizzVox is a Web-based application that lets you upload photos and talk about them. You can continue annotating each time you watch, adding more information, improving the voice record incrementally as you think of new things to say. Others can chime in if the event involves several people or family. In the end, VizzVox lets you save the story to your e-memory as a video.

Video is fantastic, but shooting it can take you out of an event, and it can be very time-consuming to edit. The good news is that editing has gotten much easier, thanks to programs like Microsoft's Movie Maker and Apple's iMovie. Both are free applications that allow you to create very professional looking movies by just dragging and dropping photographs, video, music, and voice-overs.

Whenever you can afford the time to do a really complete video shoot and production, go for it. For the rest of the time, develop the discipline of shooting video cliplets of about ten seconds each. Then create stories of no more than ten minutes. These boundaries will not only keep the story more interesting for viewers, but will provide realistic targets so that you will actually create more video stories.

Another great way to tell a story is in a scrapbook. Gather photos along with images of stuff you have digitized (tickets, dried flowers, recipes, and so on) and add captions to tell your story. There are a host of Web sites that provide electronic layouts, instructions, and even classes on how to create your e-memory in the form of a scrapbook.

Time lines are wonderful for visualizing a series of events. I can see from research projects, including MyLifeBits, that some incredibly compelling time line software will be coming to market in the coming years. The present offerings aren't too bad either. Try out www.smartdraw.com, www.timelinemaker.com, and other time line applications on the Web.

As you think about leaving your stories to future generations, don't forget the digital equivalent of cemeteries and libraries. For a fee, Web sites like www.legacy.com and www.forevernetwork.com offer to store letters, essays, photos, videos, and stories to pass on to future generations. I like the way famento.com hosts a person or family's content, because the format is a compelling timeline of photos and videos.

Everyone has an urge to tell his stories. Go ahead and tell yours; it isn't about being perfect.

STEP III: USE ALL THAT INFORMATION

You will need to organize your e-memories to get the most out of them. It turns out that the bigger the collection, the more care you need to take in how it is organized. People have learned that the way you organized a single bookshelf wasn't adequate for organizing a whole roomful of bookshelves. And the approach to a roomful of books was not enough to manage the Library of Congress. Amazing as it sounds, your e-memory will be more akin to the Library of Congress than anyone's personal library. You will be dealing with vast quantities of information.

Of course, you won't have a paid staff to take care of your e-memories like the Library of Congress does, so something has to be done to reduce the workload. Thankfully, computers are our servants. If you make sure they have good data to start with, then they can do all kinds of automatic classification and lookup. This is why I stressed having the time set correctly for your photos and adding geolocation. With those values set, the computer can sort all your photos by time and space, letting you query by place names, or loading all photos taken during a certain event. Likewise, if you perform character recognition on your scanned paper, then you can search for text in the paper.

Born-digital items tend to have information like this that is useful for organizing your collection. For instance, Web pages have the URL they came from and the date you visited them. E-mails have the address of the sender, the subject, and the date. Digital music has the name of the song, the artist, the album, and more.

Unfortunately, there is still some work for you to do organizing your e-memories. For example, if you simply scan a picture and say nothing about it, all the machine knows is that it has an image scanned on a certain date. It is up to you to add the information to make that picture useful. Also, it can be very worthwhile for you to label bills that you download from the Web. Sure, you can always search for "AT&T" and find all the documents that contain that text, but after years of receiving phone bills you will just get a big pile of them. You might then sift through them by the creation date of the file, but file creation date is notoriously unreliable (sometimes moving a file changes it). You could add the year to your search, say "AT&T 2006," to narrow it down, but you are still going to have to look in a number of files—possibly some notices from AT&T in addition to bills from that year—to find the one you want.

Suppose, instead, that your files have names like "AT&T bill 2006-08-12.pdf." Now a glance at the file name will tell you the right one. When you are downloading your bill once a month, it only takes a couple of seconds to give the file name. Or, if you are unwilling even to type a name once a month, suppose you created a folder each year like so: "2008/bills/phone/AT&T"—and then just selected that folder to save your bill in each month. Then, at least, you would know you are only looking at actual bills from AT&T of that year.

You have three ways to add information to help organize your e-memory: putting your files in a good folder structure, creating a useful file name, and adding attributes to the file. All this takes work, and there will be a bigger payoff for some items than others—so you might as well be selective about what you organize. Still, for many items you will be very glad you did a little work to keep things organized so that you can find them again. Our friend Professor William Jones, of the University of Washington, calls this activity "keeping found things found."

Your computer file-and-folder system is a sort of electronic file cabinet. Using it is no harder than figuring out what hanging folders go in your steel file cabinet and what manila files go in each hanging folder.

You may want to organize folders the way Apple or Windows suggests, which is by the type of information. They create folders like My Books, My Music, My Notes, My Pictures, My Videos, and so on. Or, you may want to create a hierarchy organized around aspects of your life, with folders like My Health, My School, My Work, and My Children. Either will serve; it's up to you. I like to organize first by the type of information, and then use an aspect of my life as a subcategory. So, I end up with the folder "My Pictures/ My Children/Brigham" for pictures of my son and his family.

Having as much information as I do stored on my computer, combined with being in my seventies, affords me with a new view, what I call "life lines." Looking back at my own life, I can see distinct chapters or well-defined segments of time: my childhood, the different schools I attended, organizations I worked for, activities, projects I carried out within each of the lives, vacations, my family, and so on. What's nice about this is that I ended up with everything about a particular life line in one place—a folder with files of pictures, correspondence, notes, and anything else related to that period.

In the Annotated References and Resources, I've included my file-and-folder hierarchy as an example of a design of this size. It's just an example, not a recommendation. Once you get your own collection going, different organizational ideas will emerge. There is no perfect structure! So, just get started and do your best.

For file names, I recommend cramming in as much information as you can. Files get moved around, so it is risky to rely on the name of the folder they are in. The more description, the better. I name my photos with the following information: What/Who; Event; Location; Date—in essence who, what, where, when. Say I'm looking at a photograph of myself and my granddaughter, Kolbe, at her eighth-grade prom. The photo would have the following name:

Bell, Gordon, and Schultz, Kolbe; Eighth-Grade Prom; Hillsborough, New Jersey; 2008-11-15

The last number is the date, November 15, 2008. I used the format year-month-day, e.g., 2008-11-15, because that makes an alphabetic sort the same as a chronological one (and when it comes to file names, alphabetic sort is all you get).

In addition to file name and folder, with some files you can add extra attributes. For example, in Microsoft Office documents, you can set the author and title of the document. For music, you can add your own rating of each song. Photos, of course, have date and location, and this is true for your scanned photos as well as your born-digital ones. The more dates and locations you set in your scanned photos, the better off you will be.

For photos, and some other file types, you can add "tags." A tag is just a word or phrase that is attached to the file. You can then search and sort by tags. Note that tags, unlike folders, are actually part of the file, so you keep them when the file is moved. Furthermore, with tags you have more freedom than with folders. A file can only be in one folder, but you can apply many tags. So, while a folder system makes you choose whether to file a photo under "My Photos/My Children" or "My Photos/Birthday parties," there is no problem adding both tags "My Children" and "Birthday parties" to a photo.

STORAGE SOUNDS SIMPLE. . . .

Learn to be aware of where your data is actually stored. Is your e-mail archive on your hard drive or on the Google e-mail server or somewhere else in the Internet? Make sure you use an e-mail client and save copies of all e-mails in your PC. You never know when that provider many decide the business is not profitable enough and close their digital doors, leaving you cut off from your e-mail. Even if you don't like using a client for your day-to-day use, use it to make backups of your e-mail. I think you will come to appreciate a client, because having the message at your fingertips without waiting for a download saves time and cloud storage space, especially when the message contains large attachments. My own mail archive is approaching ten gigabytes.

I have more than a hundred thousand Web pages saved in my e-memory by MyLifeBits. MyLifeBits Web page capture is completely automatic, and makes a copy of the actual page in addition to recording the page's address (URL). Unfortunately, at the time of writing, software to make a copy of each Web page is not commercially available. Google's toolbar has a Web history feature that records URLs and lets you search for text on pages in your history, but they won't let you download the data, so you are at their mercy to retain your e-memory. Most browsers also have a history of URLs you have visited, but what you can do with them is pretty limited. I'm amazed that there isn't something better on the market for this now and would be surprised if there isn't a good product out by the time you read this.

Without automatic recording, it is just not realistic to save every Web page you visit as I do. But there is no doubt you will save some; in particular, your bills and statements that come in HTML format. You can select "Save As" and make a copy of the page, either as a collection of files—the HTML of the page itself plus all the images and other files needed to fully display the page—or as an MHT file that wraps up all these files with the HTML file into a single file. Given the choice of these two, I'd recommend the latter as more manageable.

However, quite a few Web pages actually involve a lot of fancy programming that allows you to see what you see. When you open a saved version of some of these pages, you might find that parts of it are broken, as the browser is expecting you to be logged in using some specific Web service. To make sure that I capture what I see in the browser, I save a print version of the page. The idea is to tell the browser that you are printing out the page, but actually save the print version to your e-memory. If you have OneNote, you will see a "printer" that actually delivers the printed page into a OneNote page. Another way to do this is to install the CutePDF "printer," a free plug-in program. When you select this "printer" it lets you save what would have been printed as a PDF file. I'd recommend using a print copy of your Web pages rather than HTML, especially for your bills and statements.

You also should be logging all your text messages. Your computer chat program ought to have this as a feature; for example, with Windows Live Messenger, I simply use the setting "Automatically keep a history of my conversations." Getting SMS messages off your cell phone might require a little more effort. If you have a smartphone, you can get programs like SMS Exporter or SMS Cool!

It is really a shame that more and more communication is being buried inside social Web sites like Facebook. I don't believe you will want to make the effort presently required to save every single communication in your social Web sites. However, you should be sure to save some favorites, and occasionally just grab the look of your home page for your e-memory. Hopefully these sites will wise up soon and release our data from captivity.

I spoke above about receiving all of your bills and statements electronically. With financial transactions, you should take this even further. Rather than just converting the paper statements to electronic documents, you should also capture the information on a transaction-by-transaction basis by using a program like Quicken or Microsoft Money. These programs will download transactions from your bank, and let you add your own notes to them. You can sort, report, and search through the transactions, which is a powerful form of Total Recall. Doing online banking coupled to Quicken or Money for all your financial transactions will pay in time for everything else.

Credit card transactions are especially revealing. They are a reflection of your life. They tell you how much you spend for essentials, education, dining, entertainment, travel, favorite foods, and so on. The transaction often indicates the location where it was made, too. I can recall who that plumber or electrician was, what restaurant we went to when we had that great cassoulet, and what I got Sheridan for her birthday last year.

Your bank probably avoids sending some messages via e-mail due to fraud concerns. Instead, they have their own proprietary e-mail system that you access through their Web site. If your bank communicates with you this way, make a copy of their messages to you. You may need to explicitly cut and paste the correspondence to a document that serves as a log of correspondence.

THE BEST BACKUP

Many of us make regular backups to an external hard drive. That's a good start, but if your house burns down you could lose your both PC and your backup drive. To be really safe, you want a backup that is geographically separated from your main machines so that a natural disaster like a flood or earthquake doesn't take out everything at once.

One way to do this is to back up everything to DVDs or onto an external hard drive. Then mail the backup to a trusted friend or family member who lives far away. That forms your archive, and will be most of your e-memory, since it represents your life to date. Back up subsequent material via an Internet service such as carbonite.com, backup.com, idrive.com, ibackup.com, or one of many other such sites.

Don't think of any replica of your data as reliable. A cloud service could lose your data just as easily as your own hard drive could crash. So, before you decommission your PC, make another full backup, so that you always have at least two copies of your data.

Wells Fargo's vSafe offers a personal online "safe" to protect copies of your family's documents like birth certificates, immunization records, wills, tax documents, and more. The service can be looked at as the equivalent of a safe deposit box in a bank.

I mentioned doing quick backups to a USB drive. Another way I make sure I don't lose my most current work is by using file synchronization. I use Windows Live Sync, but you could try a competing product. Whenever I am connected to the network, Live Sync is synchronizing files in certain folders with Vicki's computer. As long as I use one of my Live Sync folders, then I am always replicated as recently as the last time I connected to the Internet. Jim Gemmell and I shared a Live Sync folder for this book, and also sent chapters in e-mails to preserve extra backup copies of them.

The real answer to what is the best backup is: have more than one.

STAY CURRENT, STAY GOLDEN

Some care is needed for file formats to make sure you can still read all your files fifty years from now. In order to avoid the Dear Appy scenario, follow three guidelines.

First, regularly convert your files to the latest formats. For example, suppose a new standard for photos called JPEG2015 takes the world by storm. Convert all your all JPEG photos to this new standard.

Second, whenever possible use "golden" formats that you believe will be supported "nearly forever." A good hint of such a format is that it is used by millions of people millions of times. Even if such a format ever became obsolete, the large market for them would guarantee that solutions would be provided to convert the old format to something new. Good examples of golden formats are JPEG, MPEG-2, HTML, and PDF.

Third, make a print version of your interactive data. For example, every year I print out an annual report from Microsoft Money to PDF. That way, no matter what happens with the Money format files in the future, I have a version that I can at least look at and search for text.

PROTECT YOUR PRIVACY

As discussed in Chapter 8, everyone needs to be concerned about the privacy and security of his or her e-memories. As we wrote this book, Jim Gemmell had someone use his credit card number for a Las Vegas spending spree. I have found some solutions that work very well.

As with backups, I believe in layers of security to protect my privacy. Invest in a firewall as a first line of defense. Nearly all the home routers have a built-in firewall.

As a second layer, make your computer secure. Make sure it has all the latest patches and is running good security software. Protect access to your computer by use of a strong password (do a search on the Web to learn what constitutes a strong password). Make your notebook require the password whenever it is started up, so that if you lose it someone can't just flip it open and get your stuff. The same goes for your smartphone—always password-protect it. Get a smartphone that also lets you remotely wipe the data if someone steals it.

As a third layer of defense, encrypt the data on your hard drive. That way, if someone pulls out your drive and puts it in another computer, he will not be able to read the contents. If you are Windows user, get a version of Windows that supports BitLocker, which implements this kind of drive encryption. Use this encryption for your external backup drive too.

ENJOY IT ALL TOO

Once you are lifelogging, I suggest buying some extra equipment to get the fullest enjoyment. Displaying your photographs and video cliplets throughout your home can be immensely rewarding, as a growing number of people are finding. You see something wonderful every day instead of waiting months to get the inclination to dust off the old memories.

If you don't have a large-screen high-definition TV, buy one, and connect it to a device that can display. If you have Windows Media Center for one of your PCs, you can display media on the TV from the PC using an "extender" such as the Linksys DMA2100, or an Xbox 360. You can also buy LCD screens from HP with built-in extenders. If you are an Apple user, get an Apple TV.

Also, add some small-screen versions of the same thing, such as Samsung's Photo Frame series. These little picture frames wirelessly connect to your home network to spread your e-memories all over your home, from your end table and kitchen counter.

With these three steps, you are well on your way to living in the Total Recall world. But it is just the beginning. In the coming years Total Recall will get much better as increased storage, better software, and a proliferation of sensor hardware makes the picture complete. Which brings me to the final step. . . .

STEP IV: BECOME AN ENTREPRENEUR (OPTIONAL)

The steps above will get you started with Total Recall today, but there is a lot that can be done to make it easier for people to participate in the Total Recall revolution. Total Recall holds great opportunities for entrepreneurs to serve the public—and make a lot of money.

Hardly a week goes by that I don't hear a pitch from some start-up that wants me as an angel investor or as a board member. Very few win me over. Below is my top ten list of Total Recall ideas I'd *like* to hear someone pitch to me.

START-UP #10—PICTURE-TAKING MIRROR

Have you ever seen one of those photo exhibits in which someone has taken a picture of his kids in a set pose over a sequence of years so that you can see them change over time? One fellow, Richard Hanson, has gone even further and taken a picture of his son every single day of his life. The result is utterly compelling, even for us strangers. Imagine if it were of yourself or a loved one. I wish I had a mirror in my hall with a camera behind it that would take my picture every day. Actually, I'd like to give one to each of my grandchildren. An alternative to the mirror would be to put the camera in one of those digital picture frames—then you'd also have a potential videophone.

START-UP #9—CONVINCING CHAT BOT
FROM MY HISTORY

I've already invested in MyCyberTwin.com, so you know I like this idea. The company's chat bots are sitting in servers and providing help services to thousands of customers a day with questions about banking, tax forms, and other service requests. MyCyberTwin is great, but we are still waiting for the first company that can take a heap of someone's correspondence (e-mail, chats, letters, et cetera) and produce a really convincing impersonation. Any team that can take my corpus and turn it into my digitally immortal chatting self will get my support. And that's not just vanity—if you can imitate me, you can imitate help-desk personnel and make a ton of money.

START-UP #8—DOCUMENT MANAGEMENT

It sounds great to declutter your life by scanning all your documents, but full-text search on a heap of files is not always the best way to retrieve information. This service (or program that you run) will automatically group similar items. It will build a knowledge base of every kind of document it can learn about, for example from all major utility and phone companies. It will be able to pull out the date, the total, and who the bill is from. It will create descriptive file names for all your documents and also create a human-readable XML file containing all the information it was able to extract. You can send in a box of documents and get back files with descriptive names in meaningful folders. For example, a file called "AT&T Bill 2008-09-17 total 87.23" in the "AT&T" folder, which is in the "Bills" folder. An accompanying file will include things like the address the bill was sent to and the breakdown of long-distance and local charges.

START-UP #7—UNIFIED STORAGE

Microsoft took at stab at unifying all my storage with the WinFS project, but gave up. The idea is to end different data types living in different "silos," like e-mail in my e-mail app, photos in my photo app, music in my music app, and so on. Instead, I can deal with them all at once and lump some e-mails together with some spreadsheets, or quickly jump from a photo to documents created on the same date. Also, I should be able to add comments and tags to everything, not just a few special types.

Short of a major operating system really solving this, there may be a place for some smaller start-up to hatch an idea for bringing together parts of my scattered data—say, grouping my e-mail with my files, or perhaps my Facebook entries with my chat logs. I'm not sure how such a smaller start-up can succeed in this space, but someone else might have a better imagination for it.

START-UP #6—TRAVELOGUE CAMERA

I want something like the SenseCam to take on vacation. It needs to be smaller and more attractive than the SenseCam prototypes I use now. In fact, it ought to become a desirable fashion accessory. It must include GPS and a microphone for voice annotation. It could be a device that talks to my cell phone and uses the cell phone's storage. Just as important as the hardware is summarization software. I want to go camping with my son's family for a week. He and I each wear one. At the end, our trip is automatically summarized and photo albums are made. We can view animated maps of our travels. Great photos taken by others of the same place that are publicly available (say, from Flickr) may be included. In short, wear one of these and get a sharp-looking travelogue with zero effort at the end of your trip.

START-UP #5—WEARABLE HEALTH
MONITORING DEVICES

This will actually be a whole class of start-ups. The typical pitch should go like this: We have device X. It is easy to wear, as either part of your clothing, or as a little wristband, or the like. It sends all its data to your cell phone, where an application stores the data and then forwards it either to your PC or to some service in the Internet that logs it forever. It must be easy to recharge. Wear it and forget about it. You'll get notified when something of interest comes up. Important events might automatically be forwarded to your doctor. The BodyBugg is a forerunner in this area that will include fitness monitors, pacemakers, and hopefully new in-body devices.

START-UP #4—CELL PHONE LOGGER

As I've said, a smart cell phone is a killer device for Total Recall. My cell phone should record my GPS location and call log, record all text messages, support note taking and dear-diary entries—both text and audio, and even video—and store it in the cloud. I'm already invested in reQall, which is a great step in cell-phone-based memory creation. Evernote also looks hot. We created some prototype software in this space and it wasn't too hard to get to a proof of concept. I expect to see some strong, more comprehensive efforts soon.

START-UP #3—DEAR APPY FORMAT
CONVERTING SERVICE

Perhaps it could be called Format Master or Yours Forever. The idea is to convert all your files that are in formats that are stale or are fringe formats in danger of becoming stale. Files would be converted to the latest format. In addition, some "just-in-case versions" could be created, such as PDF print versions of spreadsheets and Web pages. This could be a service that you upload your data to, or you might run a program on your PC. The service should be provided to storage providers. That is, suppose storage.com stores files for you. They would contract Format Master to keep your files up to date if you pay them an extra three dollars a year.

START-UP #2—DIGITAL AFTERLIFE

What happens to my bits when I die? I need a contract to store my data for two hundred years (including the Format Master service). I should be able to put some information in a time capsule, for example, not releasing the material to my family for twenty years, and only releasing to the public after one hundred years. Furthermore, for those who haven't been practicing Total Recall, we will want to send a box of their stuff (photos, documents, et cetera) to a service that will scan it all and put it in this digital cemetery. There are already some companies doing memorials and promising storage, but I still see room for a really innovative company to take this to new heights.

START-UP #1—SWISS DATA BANK

Storage must be safe from hackers, safely backed up, and plausibly deniable. It would be preferable if the data bank only ever saw my encrypted bits, so it couldn't divulge what it has, even if it wanted to. There will be specialty Swiss data banks, such as a health data bank or financial records data bank (like HealthVault and Wells Fargo, respectively, but with deniability). Whoever can build the most trusted brand name will reap big rewards. Perception will be as important as the technology; the slightest doubts about your brand could kill the business. A variant on the concept would set up peer-to-peer encrypted storage to virtually eliminate storage costs for the provider, while adding another layer between governments and your data.

CHAPTER 10

THE FUTURE

If the world follows my lead, Total Recall will be a very private matter. Encryption will be universal, e-memories will reside in Swiss data banks, and sharing will be careful and limited. I think the younger generation ought to eventually see their casual approach to privacy as a mistake and scale back their public disclosures. But maybe they won't. Maybe my attitudes regarding privacy are headed toward extinction. There are those who say privacy is gone forever and good riddance to it.

If lifelogging becomes life-blogging, then the successors to Facebook and Twitter could have detailed records of every parameter of your life, with location, biometrics, sights, and sounds. Imagine for a moment that all memories are shared. One could then dream of data-mining all these memories, looking for collective good, much the way that my personal memories will be mined for my own good. There may even arise some kind of cyber communism that demands all of your information for the public good—for example your location, to help with city planning and emergency management. There could be an appeal to your own advantage: Just as Amazon.com and other Web sites track sales to predict items that you might want to buy, the collective cyber-mind might suggest many activities, places, and things that would be to your benefit or liking.

I don't believe it. Embracing complete openness is like resched-uling Judgment Day for today. "What you did in secret will be shouted from the rooftops" might as well be the name of the next social networking Web site, echoing the words of Christ. But who can say for sure? I think that the future more than ten years from now is very hard, even impossible, to predict.

For one thing, it is simply hard to wrap one's mind around the distant future. Passing on my e-memories to my grandchildren would be exciting enough—who can digest the idea of centuries-ful of e-memories? You have thousands, if not millions, of ances-tors from the past thousand years. What if you had all of their e-memories? Would one's own family tree attract more attention than the History Channel? Would my family have a top-ten an-cestor list, and a family highlight reel? Doubtless medical history would be pored over and different conditions identified in differ-ent branches. Genes would be compared to find ancestors similar to me, and lessons would be drawn from their lifestyle and health results. I can imagine drawing inspiration from an ancestor with similar interests to mine. I can also imagine angst over a tragic ancestor with some resemblance to me. I'd talk to his cyber twin: "But why did you want that?" "Did you realize . . . ?" If some great-great-grandsire had a gap in his e-memories, I might try to get access to the memories of his friends and relatives to try to piece together what he was keeping secrets about.

But with so many possible changes in society over a thousand years, my speculations may not be much better than a wild guess.

And if culture is hard to predict very far ahead, I think technology is equally hard to predict in the long term. I don't think anyone can predict technology more than a couple of decades ahead, because that implies knowledge of materials or phenomena that have yet to be discovered. Carver Mead, a Caltech computer scientist who coined the term *Moore's Law,* posits an eleven-year rule: It takes eleven years to bring a high-tech product from the lab into existence. I feel comfortable predicting the progress of Total Recall about ten years out, based on technology that someone is already working on in some lab.

Next, I'm going to outline the technological context that will usher in Total Recall in the coming decade. There is a clear direction for computer hardware, sensors, and networking. Also, I see a trend in unified communications and storage. Total Recall will fuel demand for improved computer-user interfaces; it will also be key to enabling these interfaces to become more natural. Total Recall is far beyond speculation and, in many areas, is past the research phase and ready for straightforward product development. There is a clear path for hardware to build, software to write, and new services to develop. Product development for Total Recall has already begun in dozens of companies.

SHRINKING E-MEMORY MACHINES

The full spectrum of modern hardware is bringing us to the dawn of an era in which nearly every bit of information about your life can be captured and stored forever. This is not to diminish software. Nathan Myhrvold, the former chief technology officer at Microsoft who went on to found Intellectual Ventures, believes that only the software creator's imagination limits what hardware can provide, and he is right. But if you can't acquire it or store it, you won't be computing it.

Throughout my adult life, hardware has been rapidly changing, and it continues to do so. Moore's Law predicts that computers will be twice as powerful two years from now without changes in their size or price. Some time ago, I observed that there is another consequence of increased power and miniaturization: We get the same power at reduced size and cost. These cheaper, smaller versions of what we had before eventually get cheap enough and small enough to inspire the creation of an entirely new class of computers. A new class can be expected about every decade or so, with its own unique hardware-software environment, applications, user base, and vendors.

Eventually the power of your old PC finds its way into smaller devices, such as your digital camera, personal digital assistant, or cell phone. Looking ahead, it's easy to see lots of multifunction pocket-size devices, with vast inexpensive storage for capturing everything you see and hear at higher and higher fidelity. The cell phones of the coming decade will have enough data storage and computing power to do some very powerful calculation and data mining on you and your environment. In fact, the smartphone I presently own is about a hundred times more powerful than the minicomputers I used to design—minicomputers that were shared among entire departments.

As noted earlier, the Total Recall revolution is being birthed on the strength of a few key devices: cell phones, digital cameras, and pocket-size GPS units. These devices have us snapping pictures, sharing media, and plugging into the networked world. Your PC is still extremely useful—it will not go away—but smaller, cheaper, more comfortable, and less obtrusive devices will provide the impetus for dramatic progress toward Total Recall.

In the next ten years our pocket-size devices will be accompanied by a host of even smaller cousins that will be able to compute, communicate, and, most significantly for our purposes, sense. There is no limit to the things it might be useful to sense in timely fashion. I have already discussed some of the wonderful prospects for health sensing, sensing your location, and automatically capturing the sounds and sights of your experiences.

Sensors on and in you will know not only about your body, but your environment: the location, temperature, humidity, sound levels, proximity to wireless devices, amount of light, and air quality.

Conference rooms and home offices are likely to end up with audio and video sensing, especially as teleconferencing continues to grow. This sensing enables the capture of individuals in organizational settings.

Every appliance will be sensing and logging. For instance, your dishwasher will measure the temperature of its hot water connection, while your washing machine will know the level of vibration during each load. These features will begin as diagnostic aids for repairing individual devices, but will find themselves being used in aggregate also. When you blow a fuse and wonder what caused the overload, you will check the appliance logs to find out everything turned on at the time. Eventually, your home's e-memory will be part of your own Total Recall picture. Your time management software will be able to factor in how much laundry you do, and your health software will know you've been hauling around that heavy vacuum a lot.

You will literally sprinkle sensors in the dirt of your garden, and they will relay information through each other to a little powered hub that will forward information to your home network about soil conditions.

Your car will have its own lifebits, complete with location, health, and record of the environment it was in. It will know that

on Tuesday it was driven up a 15 percent grade in the snow, carry-
ing a load of 470 pounds and averaging 3,100 rpm. That, too, will
start as a repair diagnostic and will eventually be used to supple-
ment your own history with the story of all your driving.

NETWORKS OF UNIFIED COMMUNICATIONS

Network capacity and speed are ever growing, allowing us to move
around bigger files and watch better quality video. I'm frustrated
the television/telephone duopoly in the United States is so slow to
get us high-speed fiber-optic networks to every home. Sometimes
I wish that networking was considered part of the nation's infra-
structure, like highways, so that we could mandate fiber everywhere as in other countries. Still, the trend is in the right direction,
and we already have a pretty good start.

Total Recall will come about within the context of networks
within networks, interconnecting everything from in-body networks to home networks to global networks and finally to networks that include satellites and space vehicles. Dust-size sensors
will automatically form wireless networks and connect to everything that can be sensed. In-body implants will communicate with
each other to form a "body-area" network. The body network will
connect to the car network while you're driving. The garden network will connect to the home network. The car and home network will connect to the worldwide Internet.

This vast network of networks will host huge farms of servers
with millions of processors and many petabytes of storage space.
These farms will offer up computing and storage service to those
who need it—and, amazingly, you will need it, even though your
cell phone will boast more power and storage than your PC does
today. From the microscopic to the heavens, all will be sensed, net-

worked, and stored. This is not a forty-year-out wild guess. This is a decade-out sure bet. And I don't lose many bets.

Microsoft has pretty good communications for its employees. If you phone my work number and leave a voice mail, I get it as an audio attachment in my e-mail. Actually, if you call when I am out of the office, I may well answer because the call is forwarded to my notebook PC wherever I am, to answer using my notebook's microphone and speaker. From e-mail I can launch chat; from chat I can launch e-mail. All my chats are logged into a folder in my e-mail. All the RSS news feeds that I want to read appear in my e-mail client. I can set up my e-mail client to manage all my different e-mail accounts. I can also phone in to check my e-mail and hear it read to me.

This is called unified communications. Instead of telephone, chat, RSS, and my several e-mail accounts being completely independent, they are unified. I don't have to go around checking in different places for messages, and it's not a big hassle to switch to some other form of sending a message. Unfortunately, it seems that every new networking application out there wants to fragment my communications. My doctor makes me visit his Web site to check for messages from him. My bank sends me messages to tell me I have a message to read and reply to at their Web site. Facebook sends me an e-mail with an actual message from a friend, but makes me use their site to reply. I get a steady stream of messages from LinkedIn demanding I go over to their site to follow up. And there is a steady stream of new "must-have" communication channels. Frankly, no matter how wonderful they are, I am simply running out of time to keep up with all these virtual post office boxes.

If it is clear that communication ought to be unified rather than fragmented, it is equally clear that storage should be unified. I don't want to have to jump around between online services or computers to get my stuff. I don't want to be bothered with thinking about which hard drive on my computer contains what. And I sure don't want to be launching different applications to get at different kinds of stuff and to face all kinds of hassles in grouping dissimilar items together. Unifying storage might mean actually bringing all the information together, or it may be sufficient to just have a catalog that virtually unifies everything; to look at it as one whole while retrieving particular items from many possible sources.

The consumer demand for unified communication and storage is clear. In the long run, those who don't make it possible to unify their data or stream of communication will wither away. The number one requirement for unification is open systems that employ standards for information exchange. For example, I can get an invitation from Evite into my e-mail client's calendar because it uses a standardized calendar event format.

Once you have open systems that use standard formats, the next step is the ability to translate. Remember the health data formats that use the code "DPT" to mean different things? Translation software is required to preserve the correct meaning between systems. As anyone who has translated between languages knows, a word-for-word translation is inadequate; it gives us translations that turn "The spirit is willing but the flesh is weak" to "The alcohol is good but the meat is bad." Likewise, it can be difficult to translate between storage formats, and a lot of work is yet to go into this effort. The Semantic Web, which aims to standardize transmission and translation of information, is an important effort in this area.

There will also be a unification of networking in the sense that we will cease to have distinct networks for different types of data. Already we get telephone over our cable TV network and TV shows over our telephone's DSL. Eventually, we will get a digital dial tone that carries anything and everything. In our homes, we will not have TV, telephone, and computer network wiring; we will have a digital home network for everything, and our home server will record TV shows and telephone calls, while it also serves up our e-memories. Our telephones and cell phones will just become small terminals into this universal digital network.

NATURAL USER INTERFACES

Your e-memories will be a vast ocean of data for you to navigate. Software will be your guide, summarizing, data mining, and anticipating what you may need. Still, there will be times that you will want to explore some particular area that has piqued your interest, or search after a very specific item. To take over the helm of this kind of navigation means handling many different controls, some of which may be complex. I recall bringing up one MyLife-Bits interface and feeling as if I had sat down in the cockpit of a 747, with a host of knobs and switches to manipulate, and numerous gauges to examine. It was intimidating. But each gauge informs you and every knob empowers you, so getting rid of some of them in the name of simplicity could be dangerous.

I've mentioned already that software will be a personal assistant to take care of a lot of the Total Recall chores. It is also critical that software be able to communicate with you in a natural way, just as a personal assistant would. This is called a natural user interface (NUI). With a NUI, you can manipulate your vast e-memory collection in a complex way without the need for lengthy training .

The ideal natural user interface should be able to handle the way you already communicate. You should be able to type or speak in your normal language. "Show me pictures from my last trip to New York." "When was my last doctor's appointment?" "Sure," "OK," "Uhn-uh." Gestures should be understood according to your culture, like a thumbs-up for approval or a throat-slashing motion for cutting something off.

You should be able to talk to your computer, but talking isn't always best. Sometimes it's quicker to point at something. It is often easier to type with corrections than to dictate (I couldn't even compose this paragraph without a lot of changes in word choice—I could never have dictated it). There is an advantage to be gained by letting me interact with the computer in a natural way, whatever that may be, and not necessarily using speech.

A natural user interface would be very helpful for Total Recall. As it turns out, NUIs really need Total Recall even more than Total Recall needs them. A NUI would be severely hobbled if it had no memory or knowledge of you. For the interaction to be really natural, I must be able to use my own idioms and nicknames. I must be able to ask about "my sister" or the "Web page I saw last week."

A really natural user interface, just like a real personal assistant, would ask questions to clarify: "Did you mean your uncle Bob, or Bob at the office?" It would know the context of your conversation to make sense of what you say, just as a real person would. By tapping all of your e-memories, it would have even more context, knowing your preferences and your usual schedule. A natural user interface would know what terms and acronyms you use regularly, and which require more explanation. "Know your audience" is the first rule of public speaking. "Know the user" is the key to a natural man-machine interface, and Total Recall will finally make it possible.

EXTREME LONG-TERM PRESERVATION

I was asked to give the keynote address at the British Library's Digital Lives Conference while writing this book. There was a fascinating discussion about how the library of the future will preserve e-memories rather than papers. But looming behind all the technical details is the really big question: Who gets in the British Library's digital lifeboat and who gets left behind? However incredible the growth of storage continues to be, the library's storage will remain finite, and after all, they aren't interested in keeping the e-memories of everyone. They will continue to save only those of the most eminent politicians, authors, philosophers, and so on. And it is not clear to what "depth" they will be kept—is it better to have ten full lives or twenty lives at half-resolution?

This raises a question closer to home: Will my progeny one thousand years hence really be able to have a copy of all their ancestors' lives? As I pointed out, each person could have millions of ancestors in that time span, so each individual's having a full family tree of e-memories is out of the question. The cost of storage would have to be shared among all members of the family. We might even think of the cost being shared among the entire human family; each generation could share the cost of trying to preserve all previous generations.

The capacity of hard drives and other storage devices is growing. So, too, are the number of them being sold. In 1995, 89 million hard drives were sold. In 2008 more than 480 million were sold. Still, we cannot presume that the amount of storage each person can afford will endlessly grow, even at a modest rate. If population growth were fast enough, one could imagine each successive generation being able to carry forward the past. However, we may see negative population growth, as in some Western countries, and additional population growth in conditions of poverty will not support the retention of e-memories.

How to keep our ancestral memories after the end of exponential storage growth is an open problem. It may even be impossible. So, though it goes against my grain to say so, it may turn out that most lives need to have their storage cost reduced over time. Video, which takes the most space by far, stands to be trimmed down the most. This could mean deleting repetitious or boring parts, or it could mean reducing the resolution, for example converting high-definition video to YouTube quality. However, I am in realms beyond my ten-year time frame. Thousand-year preservation is a matter shrouded in uncertainty.

THE WISDOM OF SPORTSCASTERS

The San Jose Sharks hockey team has just scored to tie the game with less than a minute remaining.

"And what a great pass by Thornton!" proclaims play-by-play announcer Randy Hahn. "Looks like we are going to overtime, folks!"

In the postgame show, Randy notes that Joe Thornton has assisted in more goals than any other player in the National Hockey League, and that this is the ninth time this season that he has helped tie a game up that seemed lost. He narrates over a video clip of tonight's goal, and also another one from a similar game a few weeks past. He can recite the team's record with and without Thornton. He has a "telestrator" that electronically diagrams the position of Thornton each time he made one of his legendary passes.

Sportscasters like Randy Hahn give us a real foretaste of Total Recall, with endless statistics at their fingertips, and the ability to replay game clips or interviews. Sportscasters for auto racing possess an added insight into a life filled with sensors, which record such values for each car as track position, rpm, and speed, while logging track conditions such as temperature, humidity, air pressure, rainfall, and wind speed. They use their Total Recall to entertain and inform us. Their hard data confirms or debunks our sporting theories. With Total Recall, they develop deep insight into their sport.

Diarists also have a foretaste of Total Recall. Whether done for posterity, for better memory, or simply for catharsis, journaling has many practitioners. Mark Stewart, a software engineer from Great Britain, was inspired by reading about MyLifeBits to create what he calls MyLifeDisk. It is a hyperlinked, two-volume DVD chronicle of his life, including words, spreadsheets, photos, videos, and songs. You can explore his family tree, where he has lived, his memorabilia, his education, his career, and a complete accounting of his girlfriends. Mark's life-disk really illustrates where life stories are headed, and is so compelling that he was invited to present it to the British Library at the Digital Lives conference that I keynoted. A digital life is clearly a step forward in passing on one's story to posterity.

It isn't just about who was president or what wars were fought or even the troubles of your neighbors. It is about the substance of your autobiographical memories, from your environment to your myriad relationships. It is about your memories and how you remembered them. E-memories reveal the meaning of your life.

Of course, I've had my own foretaste of Total Recall with MyLifeBits. From the beginning, MyLifeBits was conceived as a project to understand the feasibility, cost, and value of storing everything in your life. It has largely served that purpose. Like Cathal Gurrin, who wouldn't give up his SenseCam, I'm not ready to give up any of my lifelogging. I know just how much it is worth.

I'm ready now to put my money where my mouth is and invest in start-ups that will take advantage of the e-memory revolution.

Total Recall will improve our lives and afterlives in many ways. It will shake our societies and change our cultures. We look back at the ages before the advent of writing as "prehistory." The next generation will look back on our era as pre–Total Recall.

ANNOTATED REFERENCES AND RESOURCES

This section is for people familiar with computer technology, or any reader with enough curiosity to dig deeper into the ideas behind Total Recall. Those eager to get started with Total Recall will find some useful pointers as well. In addition to citations for material in the book chapters, there are references and discussions of related topics that haven't been covered yet. Here you can find references to printed publications, Web sites, people, products, conferences, and research labs.

The section is arranged by chapter, and the order of material follows the order of the chapter as much as possible.

I. THE VISION

Ray Ozzie is quoted from personal correspondence with the Authors. Other references on cloud computing:

Hayes, B. 2008. "Cloud Computing." ACM, *Communications of the ACM* 51, Issue 7 (July).

Gruman, Galen, and Eric Knorr. 2008. "What Cloud Computing Really Means. *InfoWorld* (April 7).

Martin, Richard, and J. Nicholas Hoover. 2008. "Guide to Cloud Computing." *Information Week* (June 21).

Amazon Elastic Compute Cloud (Amazon EC2). http://aws.amazon.com/ec2
Microsoft Azure Services Platform. http://www.microsoft.com/azure

Science fiction that grapples with e-memories:

Sawyer, Robert J. 2003. *Hominids*. New York: Macmillan.
Halperin, James. 2000. *The Truth Machine*. New York: Ballantine Books.
Naim, Oscar. 2004. *The Final Cut*. Lions Gate Entertainment.
Westbrook, Robert. 2004. *The Final Cut*. New York: Penguin.

Another related sci-fi work is *The Observers,* where alien microrobots have been recording Earth's history in incredible detail. People in the book grapple with being watched and recorded. The aliens are able to copy all of the information related to a person to create a virtual person, raising the issue of digital immortality.

Williamson, S. Gill. 2006. *The Observers*. Lincoln, Neb.: iUniverse.

Don Norman suggested the Teddy life recorder. His other books on design are also well worth reading.

Norman, Donald A. 1992. *Turn Signals Are the Facial Expression of Automobiles*. Reading, Mass.: Addison-Wesley.

The Millennials, aka Generation Y:

Howe, Neil, and William Strauss. 2000. *Millennials Rising: The Next Great Generation*. New York: Random House.
Laurent, Anne. 2008. "Millennials: They're Here. They're Wired. Get Used to Them." *Tech Insider* (March 24) http://techinsider.nextgov.com
Safer, Morley. 2007. "The 'Millennials' Are Coming." *CBS 60 Minutes* (May 25). http://www.cbsnews.com/stories/2007/11/08/60minutes/main3475200.shtml
Olsen, Stefanie. 2005. "The 'Millennials' Usher in a New Era." CNET News.com. (November 18).
Sweeney, Richard. 2006. "Millennial Behaviors and Demographics." http://library2.njit.edu/staff-folders/sweeney/Millennials/Article-Millennial-Behaviors.doc

Strauss and Howe have an interesting generational theory that could play into the issue of why the Millennials seem to have a different attitude to privacy and technology.

Strauss, William, and Neil Howe. 1997. *The Fourth Turning*. New York: Broadway Books.

———. 1991. *Generations: The History of America's Future, 1584 to 2069*. New York: William Morrow and Company.

Abe Crystal's Ph.D. study found that "although all the students I observed were generally comfortable with technology, there was a large variance in technology-related expertise and knowledge." Gibbons and Foster (2007) were "surprised to find that students are on average no more proficient with computer technology than are librarians and faculty members. Some students demonstrated broad knowledge of computers and facility in using them, but others were awkward and clumsy."

Crystal, Abe. 2008. Design research for Personal Information Management systems to support undergraduate students, doctoral dissertation, University of North Carolina at Chapel Hill.

The International Technology Roadmap for Semiconductors is the single most important projection about the future of semiconductors. It encompasses all memory, including nonvolatile Flash memory, processors, and radios. The 2007 Roadmap projects continued biennial doubling of semiconductor density until 2016. One implication is that semiconductor memories will replace disks for portable computers.

International Technology Roadmap for Semiconductors Web site. http://www.itrs.net

Disk Storage Roadmaps are available from individual vendors and market intelligence firms like IDC that show nearly annual doubling of disk densities.

Arai, Masayuki. 2009. "Optical Disks Used for Long-Term Storage by 2010." *Tech-On!* (March 6).

Rydning, John, and Jeff Janukowicz. 2009. "Worldwide Hard Disk Drive Component 2008–2012 Forecast Update." IDC (February 1).

There is a large research community advancing work on data mining, pattern recognition, and machine learning. Here are just a few starting points:

Bishop, Christopher M. 2006. *Pattern Recognition and Machine Learning.* New York: Springer.

Kargupta, H., et al. (eds.). 2009. *Next Generation of Data Mining.* London: Chapman and Hall.

ACM SIGKDD International Conference on Knowledge Discovery and Data Mining.

International Conference on Data Mining (ICDM).

SIAM International Conference on Data Mining.

Total Recall predicts the future based on technology trends. A similar book in this genre is *Being Digital,* which did a wonderful job in 1995 of predicting our digital lives today.

Negroponte, Nicholas. 1995. *Being Digital.* New York: Alfred A. Knopf.

2. MYLIFEBITS

Million Books Project (also called the Universal Library Project) Web site. http://www.ulib.org

Project Gutenberg Web site. http://www.gutenberg.org/wiki/Main_ Page

Foer, Jonathan Safran. 2006. *Extremely Loud and Incredibly Close.* New York: Mariner Books.

Bill Gates and Jim Gray were inspirations to us.

Gates, Bill. 1996. *The Road Ahead.* New York: Penguin Books.

Gray, Jim. 1999. "What's Next? A Dozen Information-Technology Research Goals." *Journal of the ACM* 50:41–57.

About the Aaron painting program:

Cohen, Harold. 1995. "The Further Exploits of AARON, Painter." *Stanford Humanities Review* 4, issue 2 (July): Constructions of the Mind. http://www.stanford.edu/group/SHR/4-2/text/cohen.html

Memex was proposed by Bush in his *Atlantic Monthly* article.

Bush, Vannevar. 1945. "As We May Think." *Atlantic Monthly* (July). Reprinted in *Life* magazine, September 10, 1945. http://www.theatlantic.com/doc/194507/bush

This book tells you much more about Bush, his life, and his amazing technological vision.

Nyce, James M., and Paul Kahn (eds.). 1992. *From Memex to Hypertext: Vannevar Bush and the Mind's Machine.* Boston: Academic Press.

In this report, Bush proposes the National Science Foundation (NSF) and more.

Bush, Vannevar. 1945. "Science The Endless Frontier. A Report to the President by Vannevar Bush, Director of the Office of Scientific Research and Development, July 1945." Washington, D.C.: United States Government Printing Office. http://www.nsf.gov/about/history/vbush1945.htm; also available as ACLS Humanities E-Book (August 1, 2008).

Some believe that Paul Otlet, not Bush, ought to get the credit for the concept of hyperlinks for his 1934 "réseau" idea.

Wright, Alex. 2008. "The Web Time Forgot." *The New York Times* (June 17).

In the 1960s, Ted Nelson took Bush's ideas and extended them to support a new paradigm for literature in a networked world. He coined the term *hypertext* and proposed ideas that are current today, like virtually including one work inside another and using micropayments.

Nelson, Theodor Holm. 1993. *Literary Machines*. Sausalito, Calif.: Mindful Press.

Nelson, Theodor Holm. 1999. "Xanalogical Structure, Needed Now More Than Ever: Parallel Documents, Deep Links to Content, Deep Versioning, and Deep Re-Use." *Computing Surveys* (ACM) 3, issue 4es (December).

Another pioneer in the 1960s who was inspired by Bush was Douglas Englebart, who founded a research lab with the goal of "augmenting human intellect." His lab developed a hypermedia groupware system called Augment (originally called NLS). Augment supported bookmarks, hyperlinks, recording of e-mail, a journal, and more.

Engelbart, Douglas C. "Augmenting Human Intellect: A Conceptual Framework. Summary Report AFOSR-3223 Under Contract AF 49(638)-1024," SRI Project 3578 for Air Force Office of Scientific Research. Menlo Park, Calif.: Stanford Research Institute, October 1962.

———. "Authorship Provisions in AUGMENT." COMPCON '84 Digest: *Proceedings of the COMPCON Conference,* San Francisco, California, February 27–March 1, 1984, 465–72.

Many others besides us have noted the inadequacy of conventional computer file systems. Here are a few representative works.

Adar, Eytan, David Karger, and Lynn Andrea Stein. "Haystack: Per-User Information Environments," *1999 Proceedings of the Conference on Information and Knowledge Management,* Kansas City, Mo., 1999, 413–422.

Dourish, Paul, Keith Edwards, Anthony LaMarca, John Lamping, Karin Petersen, Michael Salisbury, Douglas Terry, and Jim Thornton. 2000. "Extending Document Management Systems with User-Specific Active Properties." *ACM TOIS* 18, no. 2: 140–70.

Gifford, David K., Pierre Jouvelot, Mark A. Sheldon, and James W. O'Toole Jr. "Semantic File Systems." Thirteenth ACM Symposium on Operating Systems Principles, October 1991, 16–25.

Maier, David. "Care and Feeding of Your PetDB." NSF Workshop on Context-Aware Mobile Database Management (CAMM), January 24, 2002.

Olsen, Michael A. "The Design and Implementation of the Inversion File System," 1993 Winter USENIX, San Diego, California, January 25–29, 1993.

You can find nearly everything about MyLifeBits at the Web site www. MyLifeBits.com. Here are some of the key papers:

Bell, Gordon, and Jim Gemmell. 2007. "A Digital Life." *Scientific American* (March).
Gemmell, Jim, Gordon Bell, and Roger Lueder. 2006. "MyLifeBits: A Personal Database for Everything." *Communications of the ACM* 49, issue 1 (January): 88–95.
Gemmell, Jim, Gordon Bell, Roger Lueder, Steven Drucker, and Curtis Wong. "MyLifeBits: Fulfilling the Memex Vision." ACM Multimedia '02, Juan-les-Pins, France, December 1–6, 2002, 235–38.

And here are some of the best articles written about MyLifeBits:

Cherry, Steven. 2005. "Total Recall." *IEEE Spectrum* (November).
Thompson, Clive. 2006. "A Head for Detail." *Fast Company* (November).
Wilkinson, Alec. 2007. "Remember This." *The New Yorker* (May 28).

Look for the latest on SenseCam at research.microsoft.com—here are some published papers about it:

Berry, E., N. Kapur, L. Williams, S. Hodges, P. Watson, G. Smyth, J. Srinivasan, R. Smith, B. Wilson, and K. Wood. 2007. "The Use of a Wearable Camera, SenseCam, as a Pictorial Diary to Improve Autobiographical Memory in a Patient with Limbic Encephalitis." In research.microsoft. com: Psychology Press, 582–681.
Harper, R., D. Randall, N. Smyth, C. Evans, L. Heledd, and R. Moore. 2007. "Thanks for the Memory." Human-Computer Interaction Conference, Lancaster, UK, 2007.
Laursen, Lucas. A Memorable Device, *Science* 13 March 2009: 1422–23.
Sellen, A., A. Fogg, S. Hodges, and K. Wood. "Do Life-Logging Technologies Support Memory for the Past? An Experimental Study Using Sense-

Cam." Conference on Human Factors in Computing Systems, CHI '07, Irvine, California, 2007, 81–90.

Berry, E., M. Conway, C. Moulin, H. Williams, S. Hodges, L. Williams, K. Wood, and G. Smith. 2006. "Stimulating Episodic Memory: Initial Explorations Using SenseCam." *Abstracts of the Psychonomic Society, 47th Annual Meeting* 11:56–57. Oxford: Oxford University Press.

Gemmell, Jim, Lyndsay Williams, Ken Wood, Roger Lueder, and Gordon Bell. "Passive Capture and Ensuing Issues for a Personal Lifetime Store." *Proceedings of the First ACM Workshop on Continuous Archival and Retrieval of Personal Experiences* (CARPE '04), New York, October 15, 2004, 48–55.

Hodges, Steve, Lyndsay Williams, Emma Berry, Shahram Izadi, James Srinivasan, Alex Butler, Gavin Smyth, Narinder Kapur, and Ken Wood. 2006. "SenseCam: a Retrospective Memory Aid." In Dourish and A. Friday, eds., *Ubicomp 2006: Ubiquitous Computing.* Lecture Notes in Computer Science 4206: 177–193. Berlin, Heidelberg: Springer-Verlag 2006.

StartleCam is another sensor-triggered wearable camera. It uses multiple skin conductivity sensors worn on the fingers. The sensors are used to detect the startle response in the wearer, and save the recently captured images, which will presumably be of events that aroused the user's attention.

Healey, J., and R. W. Picard. "StartleCam: A Cybernetic Wearable Camera." *Proceedings of the Second International Symposium on Wearable Computing,* Pittsburgh, Pennsylvania, October 19–20, 1998.

A system called u-Photo extends the action of picture taking to also include capturing the state of devices and sensor values in the camera's view. LED tags are placed on objects to determine if they are in view. For instance, a picture of a room may also record the information that the lights are on, that the temperature is seventy-four degrees, and that a movie is playing at a certain point.

Iwamoto, Takeshi, Genta Suzuki, Shun Aoki, Naohiko Kohtake, Kazunori Takashio, and Hideyuki Tokuda. "u-Photo: A Design and Implementation of a Snapshot Based Method for Capturing Contextual

Information." Pervasive 2004 Workshop on Memory and Sharing of Experiences, Vienna, Austria, April 20, 2004.

The remote control in your home—or perhaps that you carry with you— can even be part of lifelogging.

Abe, M., Y. Morinishi, A. Maeda, M. Aoki, and H. Inagaki. 2009. "A Life Log Collector Integrated with a Remote-Controller for Enabling User Centric Services." *IEEE Transactions on Consumer Electronics* 55, no. 1.

Cathal Gurrin's Web page is http://www.computing.dcu.ie/~cgurrin and here are some of his papers about e-memories.

Doherty, A., C. Gurrin, G. Jones, and A. F. Smeaton. "Retrieval of Similar Travel Routes Using GPS Tracklog Place Names." SIGIR 2006—Conference on Research and Development on Information Retrieval, Workshop on Geographic Information Retrieval, Seattle, Washington, August 6–11, 2006.

Gurrin, C., A. F. Smeaton, D. Byrne, N. O'Hare, G. Jones, and N. O'Connor. "An Examination of a Large Visual Lifelog." AIRS 2008—Asia Information Retrieval Symposium, Harbin, China, January 16–18, 2008.

Lavelle, B., D. Byrne, C. Gurrin, A. F. Smeaton, and G. Jones. "Bluetooth Familiarity: Methods of Calculation, Applications and Limitations." MIRW 2007—Mobile Interaction with the Real World, Workshop at the MobileHCI07: 9th International Conference on Human Computer Interaction with Mobile Devices and Services, Singapore, September 9, 2007.

Lee, H., A. F. Smeaton, N. O'Connor, G. Jones, M. Blighe, D. Byrne, A. Doherty, and C. Gurrin. 2008. "Constructing a SenseCam Visual Diary as a Media Process." *Multimedia Systems Journal,* Special Issue on Canonical Processes of Media Production (in press) 14, no. 6:341–49.

Smeaton, A. F., N. O'Connor, G. Jones, G. Gaughan, H. Lee, and C. Gurrin. "SenseCam Visual Diaries Generating Memories for Life." Poster presented at the Memories for Life Colloquium 2006, British Library Conference Centre, London, UK, December 12, 2006. [BibTex] Memories For Life Web site.

We started the CARPE (Capture, Archival and Retrieval of Personal Experiences) research workshop. Many interesting papers about lifelogging were presented at these workshops.

CARPE Web page. http://www.sigmm.org/Members/jgemmell/CARPE

The Microsoft Research Digital Memories program gave funding, MyLifeBits software, and SenseCams to fourteen universities. You can find information about the projects, including the published articles.

Digital Memories program Web site. http://research.microsoft.com/en-us/collaboration/focus/cs/memex.aspx

3. THE MEETING OF E-MEMORY AND BIOMEMORY

Some books that explain how our brains remember, misremember, and forget. Daniel Schacter's seven sins of memory are: transience (the loss of memories over time), absentmindedness (forgetting due to inattentiveness, or forgetting what you meant to do), blocking (the temporary inability to recall something, like someone's name), misattribution (assigning a memory to the wrong source), suggestibility (memories planted by suggestions or leading questions), bias (using current knowledge to revise past memories), and persistence (unwanted recall of a memory).

Kandel, Eric. 2007. *In Search of Memory: The Emergence of a New Science of Mind*. New York: W. W. Norton and Co.

Schacter, Daniel L. 2001. *The Seven Sins of Memory: How the Mind Forgets and Remembers*. New York: Houghton Mifflin.

Wang, Sam, and Sandra Aamodt. 2008. *Welcome to Your Brain: Why You Lose Your Car Keys but Never Forget How to Drive and Other Puzzles of Everyday Life*. New York: Bloomsbury USA.

Forgetfulness tends to get worse as we age, and midlife is often marked by a sharp increase in absentmindedness, as delightfully described by Cathryn Jakobson Ramin in *Carved in Sand*. She takes the reader on a midlife quest for improved memory, covering the gamut from synthetic estrogen to mental aerobics.

Ramin, Cathryn Jakobson. 2007. *Carved in Sand: When Attention Fails and Memory Fades in Midlife*. New York: HarperCollins.

We quote Joseph LeDoux answering the Edge's 2008 question "What have you changed your mind about?" LeDoux is the author of *The Synaptic Self: How Our Brain Becomes Who We Are*.

LeDoux in Edge. http://www.edge.org/q2008/q08_1.html

David Allen's *Getting Things Done* advocates a "logical and trusted system outside of your head." His insight that "if your reference material doesn't have a nice clean edge to it, the line between actionable and nonactionable items will blur" points out how important an e-memory of reference material, in addition to action items, can be. In addition to the *Getting Things Done* book, there are conferences, seminars, software tools, and a host of Web articles devoted to helping you implement the *Getting Things Done* methodology.

Allen, David. 2003. *Getting Things Done: The Art of Stress-Free Productivity*. New York: Penguin Books.
Allen's Web site. http://www.davidco.com

Regarding false memories of sexual abuse, Elizabeth Loftus writes, "Now, nearly two decades later, there are hundreds of studies to support a high degree of memory distortion. People have recalled nonexistent broken glass and tape recorders, a clean-shaven man as having a mustache, straight hair as curly, and even something as large and conspicuous as a barn in a bucolic scene that contained no buildings at all." "The repressed memory cases," she observes, "are another outlet for women's rage over sexual violence. Although women's anger is certainly justified in many cases, and may be justified in some repressed memory cases too, it is time to stop and ask whether the net of rage has been cast too widely, creating a new collective nightmare."

Loftus, Elizabeth F. 1993. "The Reality of Repressed Memories." *American Psychologist* 48:518–37.

Danitz, Tiffany. 1997. "Making Up Memories?" *Insight on the News* (December 15).

Researchers at Dublin City University have done some really interesting work on dealing with thousands of SenseCam pictures (see also the work under Cathal Gurrin from the previous chapter).

Blighe, M., H. Le Borgne, N. O'Connor, A. F. Smeaton, and G. Jones. "Exploiting Context Information to Aid Landmark Detection in SenseCam Images." ECHISE 2006—2nd International Workshop on Exploiting Context Histories in Smart Environments—Infrastructures and Design, 8th International Conference of Ubiquitous Computing (Ubicomp 2006), Orange County, California, September 17–21, 2006.

Byrne, D. 2007. "SenseCam Flow Visualization for LifeLog Image Browsing." *BCS IRSG Informer*, no. 22 (Spring).

Byrne, D., B. Lavelle, A. Doherty, G. Jones, and A. F. Smeaton. "Using Bluetooth and GPS Metadata to Measure Event Similarity in SenseCam Images." IMAI '07—5th International Conference on Intelligent Multimedia and Ambient Intelligence, Salt Lake City, Utah, July 18–24, 2007.

Doherty, A., A. F. Smeaton, K. Lee, and D. Ellis. "Multimodal Segmentation of Lifelog Data." Eighth RIAO Conference—Large-Scale Semantic Access to Content (Text, Image, Video and Sound), Pittsburgh, Pennsylvania, May 30–June 1, 2007.

Lee, Hyowon, Alan F. Smeaton, Noel E. O'Connor, and Gareth J. F. Jones. "Adaptive Visual Summary of LifeLog Photos for Personal Information Management." AIR 2006—First International Workshop on Adaptive Information Retrieval, Glasgow, UK, October 14, 2006.

O'Conaire, C., N. O'Connor, A. F. Smeaton, and G. Jones. "Organizing a Daily Visual Diary Using Multi-Feature Clustering." SPIE Electronic Imaging—Multimedia Content Access: Algorithms and Systems (EI121), San Jose, California, January 28–February 1, 2007.

Smeaton, A. F. "Content vs. Context for Multimedia Semantics: The Case of SenseCam Image Structuring." SAMT 2006—*Proceedings of the First International Conference on Semantics and Digital Media Technology*. Lecture Notes in Computer Science (LNCS), Athens, Greece, December 6–8, 2006.

Smeaton, A. F., D. Diamond, and B. Smyth. "Computing and Material Sciences for LifeLogging." Presented at the Memories for Life Network Workshop 2006, British Library Conference Centre, London, UK, December 11, 2006. Memories for Life Web site.

There is lots of other work on automatic summarization; for example, this paper on video summarization:

Sundaram, Hari, and Shih-Fu Chang. "Condensing Computable Scenes Using Visual Complexity and Film Syntax Analysis." Second IEEE International Conference on Multimedia and Expo (ICME-2001), Tokyo, Japan, August 2001.

We mention a wearable system from the University of Tokyo that includes a brain wave sensor in a baseball cap. Their system also continuously captures video, GPS, gyroscope, and accelerometer data.

Aizawa, Kiyoharu, Datchakorn Tancharoen, Shinya Kawasaki, and Toshihiko Yamasaki. "Efficient Retrieval of Life Log Based on Context and Content." *Proceedings of the First ACM Workshop on Continuous Archival and Retrieval of Personal Experiences* (CARPE '04), New York, October 15, 2004, 21–31.
Hori, Tetsuro, and Kiyohara Aizawa. "Context-Based Video Retrieval System for the Life-Log Applications." *Proceedings of the 5th ACM SIGMM International Workshop on Multimedia Information Retrieval*, Berkeley, California, 2003.

reQall has a Web site, and there are a number of good articles about reQall, including the one in *Forbes*, below.

reQall Web site. http://www.reQall.com
Woyke, Elizabeth. 2008. "You Must Remember This." *Forbes* (September 30).

reQall cofounder Sunil Vemuri previously developed a PDA system to record audio and location. Calendar, e-mail, and common Web site and weather reports could also be captured. Speech-to-text was performed, and regular text search was augmented with phonetic "sounds-like" search. A speaker identification algorithm was also used, and text was colored according to speaker.

Vemuri, S., C. Schmandt, W. Bender, S. Tellex, and B. Lassey. "An Audio-Based Personal Memory Aid." In *Proceedings of Ubicomp 2004: Ubiquitous Computing*, Nottingham, UK, September 7–10, 2004, 400–17.

Supermemo:

Supermemo Web site. http://www.supermemo.com
Wolf, Gary. 2008. "Want to Remember Everything You'll Ever Learn? Surrender to This Algorithm." *Wired* (April 21).

Rank Xerox EuroParc had a number of projects involving capture in aid of human memory. *Pepys* was a diary automatically created and mailed to the user each day. It was based on location information derived from Active Badges worn by the users. Episodes were recognized (e.g., "Meeting with Joe" or "Working in office"), false positives were rejected (e.g., people sharing an office should not be presumed to be constantly in a meeting), and the level of detail was reduced to a workable summary. A "video diary" (actually a series of images captured at a rate of around ten frames per minute) was also captured by cameras in the building. *Marcel* was a paper document tracking system that used a video camera mounted over the user's desk. Documents on the desk were compared with the database of documents to identify them. *Forget-me-not* was a memory aid system. It logged e-mail, file sharing, printing, and telephone calls and supported browsing of a user's diary, with events filtered by when, where, or who to aid recall. A study compared three conditions: (1) no computer support, (2) *Pepys*, (3) "video diary." They found *Pepys* improved recall, and video did even more. People and objects were the most common memory cues.

Eldridge, Margery, Michael Lamming, and Mike Flynn. 1992. "Does a Video Diary Help Recall?" In A. Monk, D. Diaper, and M. D. Harrison (eds.), *People and Computers VII*. Cambridge: Cambridge University Press, pp. 257–69, also published as Technical Report EPC-1991-124, Rank Xerox Research Center, Cambridge, UK, 1992.

Lamming, M., and M. Flynn. "Forget-me-not: Intimate Computing in Support of Human Memory," *Proceedings of FRIEND21,'94 International Symposium on Next Generation Human Interface,* Meguro Gajoen, Japan, 1994.

Lamming, M., P. Brown, K. Cater, M. Eldridge, M. Flynn, G. Louie, P. Robinson, and A. Sellen. 1994. "The Design of a Human Memory Prosthesis." *The Computer Journal 37,* no, 3:153–63.

Lamming, M. G. "Using Automatically Generated Descriptions of Human Activity to Index Multi-Media Data." IEEE Colloquium on Multimedia Communications and Applications, February 7, 1991, pp. 5/1–5/3.

Lamming, M. G., and W. M. Newman. 1992: "Activity-based Information Retrieval: Technology in Support of Personal Memory." *Personal Computers and Intelligent Systems: Information Processing '92.* Amsterdam: North-Holland, 68–81.

Newman, W. M., M. A. Eldridge, and M. G. Lamming. 1991. "Pepys: Generating Autobiographies by Automatic Tracking." *Proceedings of the Second European Conference on Computer-Supported Cooperative Work—ECSCW '91.* Dordrecht, Netherlands: Kluwer, 175–88.

The *SPECTOR* project looks at how to use *Pepys*-like automatic diaries to develop a model of the user and perform machine learning to help the user.

Kröner, Alexander, Stephan Baldes, Anthony Jameson, and Mathias Bauer. "Using an Extended Episodic Memory Within a Mobile Companion." Pervasive 2004 Workshop on Memory and Sharing of Experiences, Vienna, Austria, April 20, 2004.

The *Infinite Memory Multifunction Machine (IM³)* was a system to automatically capture every document that a user copies, faxes, or prints. It was able to automatically detect duplicates, and was quite successful in automatically filing new documents into a user's existing file hierarchy based on word counting and image analysis. A study with twenty users over two years showed that the average age of a retrieved document was forty-four days, with 10 percent of all accesses being for documents older than six months. This debunked a common conjecture that old documents would virtually never be needed.

Hull, Jonathan J., and Peter Hart. "The Infinite Memory Multifunction Machine (IM3)." *Pre-Proceedings of the Third IAPR Workshop on Document Analysis Systems,* Nagano, Japan, November 4–6, 1998, 49–58.

Hull, Jonathan J., Dar-Shyang Lee, John Cullen, and Peter Hart. "Document Analysis Techniques for the Infinite Memory Multifunction Machine." *Proceedings of the 10th International Workshop on Database and Expert Systems Applications,* Florence, Italy, September 1–3, 1999, 561–65.

Hull, Jonathan J., and Peter E. Hart. 2001. "Toward Zero Effort Personal Document Management." *IEEE Computer* 34, no. 3 (March).

Here are several works that discuss expanding our definition of mind to encompass e-memories. For instance, David Chalmers, who says in an interview with *The Philosophers' Magazine,*

> When bits of the environment are hooked up to your cognitive system in the right way, they are, in effect, part of the mind, part of the cognitive system. So, say I'm rearranging Scrabble tiles on a rack. This is very close to being analogous to the situation when I'm doing an anagram in my head. In one case the representations are out in the world, in the other case they're in here. We say doing an anagram on a rack ought be regarded as a cognitive process, a process of the mind, even though it's out there in the world.
>
> . . . A whole lot of my cognitive activities and my brain functions have now been uploaded into my iPhone. It stores a whole lot of my beliefs, phone numbers, addresses, whatever. It acts as my memory for these things. It's always there when I need it.

Baggini, Julian. "A Piece of iMe: An Interview with David Chalmers." 2008. *The Philosophers' Magazine,* Issue 43 (4th Quarter).

Clark, Andy. 2008. *Supersizing the Mind.* Introduction by David J. Chalmers. New York: Oxford University Press.

Noë, Alva. 2009. *Out of Our Heads: Why You Are Not Your Brain, and Other Lessons from the Biology of Consciousness.* New York: Hill and Wang.

Jill Price has an astonishing memory, and it seems to be a burden to her. The human mind seems to improve memory at the price of unwanted recall—this won't be the case for e-memory.

Price, Jill, and Bart Davis. 2008. *The Woman Who Can't Forget: The Extraordinary Story of Living with the Most Remarkable Memory Known to Science.* New York: The Free Press.

Stephen Wiltshire uses a photographic memory to amaze people with his art.

Adams, Stephen. 2008. "Stephen Wiltshire, the Human Camera Who Drew London from Memory." *Telegraph* (April 3).

The Stephen Wiltshire Gallery Web site. http://www.stephenwiltshire.co.uk

4. WORK

Regarding DARPA's LifeLog and ASSIST:

Archived copy of the LifeLog Proposer Information Pamphlet, http://web.archive.org/web/20030603173339/http%3a//www.darpa.mil/ipto/Solicitations/PIP_03-30.html

Safire, William. 2003. "Dear Darpa Diary." *The New York Times* (June 5).

Shachtman, Noah. 2004. "Pentagon Kills LifeLog Project." *Wired* (February 4).

———. "Pentagon Revives Memory Project." *Wired* (September 13).

The soldier patrol story is based on Thad Starner's talk at CARPE 2006, but also borrow's from other publications related to ASSIST. Special thanks to Daniel Ashbrook.

Minnen, David, Tracy Westeyn, Peter Presti, Daniel Ashbrook, and Thad Starner. "Recognizing soldier activities in the field." *Proceedings of International IEEE Workshop on Wearable and Implantable Body Sensor Networks (BSN)*, Aachen, Germany, March 2007.

Schlenoff, Craig, et al. "Overview of the First Advanced Technology Evaluations for ASSIST." *Proceedings of Performance Metrics for Intelligent Systems (PerMIS) 2006*, IEEE Press, Gaithersburg, Maryland, August 2006.

Stevers, Michelle Potts. "Utility Assessments of Soldier–Worn Sensor Systems for ASSIST." *Proceedings of the Performance Metrics for Intelligent Systems Workshop*, 2006.

Starner, Thad. "The Virtual Patrol: Capturing and Accessing Information for the Soldier in the Field." *Proceedings of the 3rd ACM Workshop on Continuous Archival and Retrieval of Personal Experiences*, Santa Barbara, California, 2006.

Glass Box:

Cowley, Paula, Jereme Haack, Rik Littlefield, and Ernest Hampson. "Glass Box: Capturing, Archiving, and Retrieving Workstation Activities." *Proceedings of the 3rd ACM Workshop on Continuous Archival and Retrieval of Personal Experiences*, Santa Barbara, California, 2006.

The Microsoft Research VIBE team, led by Mary Czerwinski, has developed a number of excellent visualizations, including time spent on your computer, and browsing your e-memories.

VIBE Web page. http://research.microsoft.com/en-us/groups/vibe
Brush, A. J., Brian Meyers, Desney Tan, and Mary Czerwinski. "Understanding Memory Triggers for Task Tracking." In *Extended Abstracts at CHI 2007 Conference on Human Factors in Computing Systems*, Association for Computing Machinery, Inc., April 2007.
Smith, Greg, Mary Czerwinski, Brian Meyers, Daniel Robbins, George Robertson, and Desney Tan. 2006. "FacetMap: A Scalable Search and Browse Visualization." In *IEEE Transactions on Visualization and Computer Graphics*.

The wearable *Remembrance Agent* is hardware consisting of a one-handed chording keyboard with a heads-up display, along with radio and infrared receivers. Location is detected using radio location beacons and people are detected using infrared name badges. It runs note-taking software that selects old notes to show you based on your current location, people it detects around you, and the text of notes that you are presently writing. A desktop version of the *Remembrance Agent* operates within a text editor and brings up relevant items in a window based on what you are typing. A contextual retrieval application called *Margin Notes* has also been developed for Web browsing.

Rhodes, Bradley. 2003. "Physical Context for Just-in-Time Information Retrieval." *IEEE Transactions on Computers* 52, no. 8 (August): 1011–14.

————. 1997. "The Wearable Remembrance Agent: A System for Augmented Memory." Special Issue on Wearable Computing, *Personal Technologies Journal* 1:218–24.

Rhodes, Bradley J. "Margin Notes: Building a Contextually Aware Associative Memory" (html), to appear in *The Proceedings of the International Conference on Intelligent User Interfaces* (IUI '00), New Orleans, Louisiana, January 9–12, 2000.

Rhodes, Bradley, and Pattie Maes. 2000. "Just-in-Time Information Retrieval Agents." Special issue on the MIT Media Laboratory, *IBM Systems Journal* 39, nos. 3 and 4: 685–704.

Rhodes, Bradley, and Thad Starner. "The Remembrance Agent: A Continuously Running Automated Information Retrieval System. *The Proceedings of the First International Conference on the Practical Application of Intelligent Agents and Multi Agent Technology* (PAAM '96), London, UK, April 1996, 487–95.

DARPA is funding work toward a vision that "a cognitive computer system should be able to learn from its experience, as well as by being advised. It should be able to explain what it was doing and why it was doing it, and to recover from mental blind alleys. It should be able to reflect on what goes wrong when an anomaly occurs, and anticipate such occurrences in the future. It should be able to reconfigure itself in response to environmental changes. And it should be able to be configured, maintained, and operated by nonexperts." CALO (Cognitive Assistant that Learns and Organizes) and RADAR are projects funded by DARPA toward this end.

CALO Web site. http://caloproject.sri.com
RADAR Web site. http://www.radar.cs.cmu.edu

DevonThink:

DevonThink Web page. http://www.devon-technologies.com/products/devonthink
Oakes, Chris. 2004. "Software Makes a Tiger of Panther." *Wired* (July 6).

DSpace is "an open-source platform for accessing, managing, and preserving scholarly works. Developed by MIT Libraries and HP Labs, DSpace preserves and enables easy and open access to all types of digital content including text, images, moving images, mpegs, and data sets in an institutional repository." "A university-based institutional repository is a set of services that a university offers to the members of its community for the management and dissemination of digital materials created by the institution and its community members. It is most essentially an organizational commitment to the stewardship of these digital materials, including long-term preservation where appropriate, as well as organization and access or distribution."

Dspace Web site. http://www.dspace.org

Naone, Erica. 2007. "DSpace Goes Olympic." *Technology Review* (November/December).

Sharing information about repairing Xerox copiers:

Bobrow, D. G., and J. Whalen. 2002. "Community Knowledge Sharing in Practice: The Eureka Story." *Reflections, the SOL Journal* 4, issue 2 (Winter): 47–59.

To be productive with your e-memories you must employ Personal Information Management (PIM). There is some great research being done on the topic.

Jones, William. 2008. *Keeping Found Things Found: The Study and Practice of Personal Information Management (Interactive Technologies)*. Burlington, Mass.: Morgan Kaufman.

Jones, William, and Jaime Teevan, eds. 2007. *Personal Information Management*. Seattle: University of Washington Press.

http://pim.ischool.washington.edu

An overview and primer about the nature and organization of information includes descriptions of the Dewey Decimal systems and facets:

Wright, Alex. 2007. *Glut: Managing Information Through the Ages*. Ithaca N.Y.: Cornell University Press.

5. HEALTH

A number of facts cited in this chapter came from presentations at the 2007 New Paradigms in Using Computers (NPUC) workshop. We used information from Dr. Paul Tang, chief medical information officer for Sutter Health, Elizabeth Mynatt of the Georgia Institute of Technology, and Peter Miller of Vanderbilt HealthTech Laboratory,

NPUC 2007 Web page. http://www.almaden.ibm.com/cs/user/npuc2007

Some other sources on electronic health records:

"VA's Electronic Patient Records Are a Model to Industry." United States Department of Veterans Affairs press release, http://www1.va.gov/opa/pressrel/pressrelease.cfm?id=1277

"Medication Errors Injure 1.5 Million People and Cost Billions of Dollars Annually; Report Offers Comprehensive Strategies for Reducing Drug-Related Mistakes." National Academies press release, July 20, 2006, http://www8.nationalacademies.org/onpinews/newsitem.aspx?RecordID=11623 (refers to the following reference)

Committee on Identifying and Preventing Medication Errors (author), Philip Aspden, Julie Wolcott, J. Lyle Bootman, Linda R. Cronenwett (eds.). 2007. *Preventing Medication Errors*. Washington, D.C.: National Academies Press.

"Health Information Technology: Can HIT Lower Costs and Improve Quality?" http://www.rand.org/pubs/research_briefs/RB9136/index1.html

Amarasingham, Ruben, MD, MBA; Laura Plantinga, ScM; Marie Diener-West, Ph.D.; Darrell J. Gaskin, Ph.D.; Neil R. Powe, MD, MPH, MBA. 2009. "Clinical Information Technologies and Inpatient Outcomes: A Multiple Hospital Study." *Archives of Internal Medicine* 169, no. 2 (January 26): 108–14.

Carter, Jerome H. 2008. *Electronic Health Records: A Guide for Clinicians and Administrators*. Second edition. Philadelphia: American College of Physicians Press.

The UK "spine":

NHC connecting for health Web site. http://www.connectingforhealth.
nhs.uk
http://www.connectingforhealth.nhs.uk/resources/systserv/spine-
factsheet
Cross, Michael. 2006. "Getting Hospital Data to Connect to the NHS
'Spine.'" *The Guardian* (August 10).

The European Union e-Health action plan:

European Commission. "The Right Prescription for Europe's eHealth."
http://ec.europa.eu/information_society/activities/health/
policy_action_plan/index_en.htm
Sherriff, Lucy. 2004. "European Healthcare 'Online by 2008.'" *The Register* (May 5).

Articles about implementing electronic health records:

McGee, Marianne Kolbasuk. 2007. "Why Progress Toward Electronic
Health Records Is Worse Than You Think." *Information Week* (May
26).
Darcé, Keith. 2007. "Unhealthy Records." *San Diego Union-Tribune* (May
20).
Merlin, Bruce. 2007. "What Killed the Santa Barbara County Care Data
Exchange?" *iHealthBeat* (March 12).

Molecular imaging agents have been developed by companies like
CellPoint:

http://cellpointweb.com

Comprehensive blood sampling by companies like BioPhysical:

http://www.biophysicalcorp.com/biomarker-research

Philips has some home health-care devices, including scales, blood pressure cuffs, pulse oximeters, and glucose meters:

http://www.medical.philips.com/us/homehealth/index.wpd

Also by Philips: "The world's first large-scale, randomized prospective telemonitoring trial showed that homebased telemonitoring reduced the number of days spent in hospital by 26% and led to an overall 10% cost savings compared to nurse telephone support. Home Telemonitoring also significantly improved survival rates relative to usual care and led to high levels of patient satisfaction."

http://www.medical.philips.com/main/products/telemonitoring/
    assets/docs/TEN-HMS_White_Paper_FINAL.pdf

More devices and approaches to biometrics:

Brady, S., S. Coyle, and D. Diamond. "Smart Shoes for Healthcare and Security." In *pHealth 2007*, Chalkidiki, Greece.

Chan, K. W., Hung, K., Zhang, Y. T. "Noninvasive and Cuffless Measurements of Blood Pressure for Telemedicine." Engineering in Medicine and Biology Society, 2001. *Proceedings of the 23rd Annual International Conference of the IEEE*, 2001.

Coyle, S., Y. Wu, K. Lau, J. H. Kim, S. Brady, G. Wallace, and D. Diamond. "Design of a Wearable Sensing Platform for Sweat Analysis." In *pHealth 2007*, Chalkidiki, Greece.

Jaimes, Alejandro. "Sit Straight (and Tell Me What I Did Today): A Human Posture Alarm and Activity Summarization System." *Proceedings of the 2nd ACM Workshop on Continuous Archival and Retrieval of Personal Experiences* (CARPE '05), November 2005.

Oliver, Nuria, and Fernando Flores-Mangas. "HealthGear: A Real-time Wearable System for Monitoring and Analyzing Physiological Signals. *Proceedings of the International Conference on Body Sensor Networks* (BSN'06). MIT, Boston, Massachusetts, April 2006.

Ooe, Yasuhiko, Kentaro Yamasaki, and Tsukasa Noma. "PWS and PHA: Posture Web Server and Posture History Archiver." *Proceedings of the*

First ACM Workshop on Continuous Archival and Retrieval of Personal Experiences (CARPE '04), New York, October 15, 2004, 99–104.

Tan Ee Lyn. 2007. "HK Invents Pain-Free Device to Measure Blood Sugar," Reuters (May 7).

Teller A., and J. I. Stivoric. "The BodyMedia Platform: Continuous Body Intelligence." *Proceedings of the 1st ACM Workshop on Continuous Archival and Retrieval of Personal Experiences,* New York, 2004.

Ya-Ti Peng, Ching-Yung Lin, and Ming-Ting Sun. "Multi-Modality Sensor for Sleep Quality Monitoring and Logging." eChronicles Workshop, ICDE, 2006.

BodyMedia Web site. http://www.bodymedia.com

"Needle-Free Blood and Tissue Measurements," http://sci.rutgers.edu/forum/archive/index.php/t-1470.html

BiacaMed Web site. http://www.biancamed.com

Omron health care Web site. http://www.omronhealthcare.com

Oregon Scientific sports and fitness products. http://www2.oregonscientific.com/

Smartex Wealthy system. http://www.smartex.it/garment_en.html

It appears that the cell phone will be the hub for a lot of health information. Allen Cheng at the University of Pittsburgh is working on HeartToGo, for mobile ECG. Daniel Fletcher at UC Berkeley converts the cell phone into a portable microscope for disease diagnosis with CellScope. Portable ultrasound is under development by Richard and Zar at Washington University.

Zhanpeng Jin, Joseph Oresko, Shimeng Huang, and Allen C. Cheng. "HeartToGo: A Smartphone-based Mobile Platform for Continuous and Real-Time Cardiovascular Disease Monitoring." In *Proceedings of the Microsoft External Research Symposium 2009* (invited paper), Redmond, Washington, March 2009.

Richard, W. D., D. M. Zar, and R. Solek. 2008. "A Low-Cost B-Mode USB Ultrasound Probe." *Ultrasonic Imaging* 30:21–28.

"Mobile phone microscopes: Doctor on call." 2008. *The Economist* (May 15).

Storing your health information:

Quicken Health Web site. http://quickenhealth.intuit.com
Microsoft HealthVault Web site. http://www.healthvault.com
Google Health Web site. http://www.google.com/health

10,000 Steps a Day:

Isaacs, Greg. 2006. *10,000 Steps a Day to Your Optimal Weight: Walk Your Way to Better Health.* Santa Monica, California: Bonus Books, Inc.

"Laura" the e-nurse:

Elton, Catherine. 2007. "'Laura' Makes Digital Health Coaching Personal." *Boston Globe* (May 21).

6. LEARNING

Deb Roy's Speechome:

Biever, Celeste. 2006. "Watch Language Grow in the 'Baby Brother' House." *New Scientist* (May 15).
Roy, Deb, et al. "The Human Speechome Project." *Proceedings of the 28th Annual Cognitive Science Conference,* 2006.

Study comparing lectures with Web-based learning and activities:

Wallace, David R., and Suzanne T. Weiner. 1998. "A Comparison of a Lecture-Style Second Coverage of Materials vs. Limited-Coverage Guided Experiential Activity. *ASEE Journal of Engineering Education.*

On the Web:

http://scholar.google.com
http://books.google.com
http://www.archive.org
http://citeseerx.ist.psu.edu
http://www.questia.com

ePortfolios:

Batson, T. 2002. "The Electronic Portfolio Boom: What's It All About?" *Syllabus* (Dec 1).

Mason, R., C. Pegler, and M. Weller. 2004. "E-Portfolios: An Assessment Tool for Online Courses." *British Journal of Educational Technology* 25, no. 6:717–27.

On tablet PCs for education, *IEEE Computer* ran a special issue on "Tablet PC Technology: The Next Generation" that included articles on "Ink, Improvisation, and Interactive Engagement: Learning with Tablets," "Classroom Presenter: Enhancing Interactive Education and Collaboration with Digital Ink," and "Facilitating Pedagogical Practices Through a Large-Scale Tablet PC Deployment."

IEEE Computer, September 2007.

The Gray paradigm of science:

Gray, Jim, Alexander S. Szalay, Ani R. Thakar, Christopher Stoughton, and Jan vandenBerg. 2002. "Online Scientific Data Curation, Publication, and Archiving." *Microsoft Technical Report MSR-TR-2002-74* (July).

The Computer History Museum has some online exhibits, as well as some transcribed oral histories.

http://www.computerhistory.org

There are many books, articles, and even think tanks on lifelong learning, for example:

Academy of Lifelong learning Web site. http://www.academy.udel.edu

Cohn, E., and J. T. Addison. 1998. "The Economic Returns to Lifelong Learning in OECD Countries." *Education Economics* 6, no. 3 (December): 253–307.

Field, John. 2006. *Lifelong Learning and the New Educational Order.* Stoke-on-Trent, Staffordshire, UK: Trentham Books.

7. PERSONAL LIFE, AND AFTERLIFE

Organic light emitting polymer (OLEP) and Organic light emitting diode (OLED) displays:

Russell, Terrence. 2008. "Samsung Gearing up for OLED Push in 2009/2010." *Wired* (April 22).

Shinar, Joseph. 2003. "Organic Light-Emitting Devices: A Survey." New York: Springer.

TripReplay from MyLifeBits is described here:

Aris, Aleks, Jim Gemmell, and Roger Lueder. 2004. "Exploiting Location and Time for Photo Search and Storytelling in MyLifeBits." Microsoft Research Technical Report MSR-TR-2004-102 (October).

Adding summarization to visualization for geolocated photos:

Ahern, Shane, Mor Naaman, Rahul Nair, Jeannie Yang. "World Explorer: Visualizing Aggregate Data from Unstructured Text in Geo-Referenced Collections." In *Proceedings, Seventh ACM/IEEE-CS Joint Conference on Digital Libraries* (JCDL 07), June 2007.

The *Stuff I've Seen* project did some experiments that showed how displaying milestones alongside a timeline may help orient the user. Horvitz et al. used statistical models to infer the probability that users will consider events to be memory landmarks.

Ringel, M., E. Cutrell, S. T. Dumais, and E. Horvitz. 2003. "Milestones in Time: The Value of Landmarks in Retrieving Information from Personal Stores." *Proceedings of IFIP Interact 2003*.

Horvitz, Eric, Susan Dumais, and Paul Koch. "Learning Predictive Models of Memory Landmarks." CogSci 2004: 26th Annual Meeting of the Cognitive Science Society, Chicago, August 2004.

Pondering digital immortality with Jim Gray back in 2001:

Bell, G., and J. N. Gray. 2001. "Digital Immortality." *Communications of the ACM* 44, no. 3 (March): 28–30.

MyCyberTwin:

MyCyberTwin Web site. www.mycybertwin.com
Roush, Wade. 2007. Your Virtual Clone. *Technology Review* (April 20).

The Turing test:

Turing, A. 1950. "Computing Machinery and Intelligence." *Mind* 59, no. 236: 433–60.

Creating biographical and family histories:

LifeBio: www.lifebio.com, formed in 2000, has a process, tools, and software to enable a person, family, or groups to create stories and documents that can be printed or displayed on the Web.

8. REVOLUTION

Dear Appy:

Bell, Gordon. 2000. "Dear Appy" *ACM Ubiquity*, 1, no. 1 (February).

Bannon argues in favor of forgetting:

Bannon, Liam. 2006. "Forgetting as a Feature, Not a Bug: The Duality of Memory and Implications for Ubiquitous Computing." *CoDesign* 2, no. 1 (March): 3–15.

Management guru Drucker on managing yourself:

Drucker, Peter. 1999. "Managing Oneself." *Harvard Business Review* (March–April): 65–74.

Google street views:

Derbyshire, David, and Arthur Martin. 2008. "Google 'Burglar's Charter' Street Cameras Given the All Clear by Privacy Watchdog." *Mail Online* (July 31).

Weeks, Carly. 2007. "Google's Detailed Streetscapes Raise Privacy Concerns. *National Post* (September 11).

"Google Street Views, Cool or Creepy?" *New York Post* (June 7, 2007).

E-memories in court:

"Electronic storage devices function as an extension of our own memory," Judge Pregerson wrote, in explaining why the government should not be allowed to inspect them without cause. "They are capable of storing our thoughts, ranging from the most whimsical to the most profound." The magistrate judge, Jerome J. Niedermeier of Federal District Court in Burlington, Vt., used an analogy from Supreme Court precedent. It is one thing to require a defendant to surrender a key to a safe and another to make him reveal its combination.

Liptak, Adam. 2008. "If Your Hard Drive Could Testify." *The New York Times* (January 7).

The idea of purposely falsifying some of your records to avoid them being used in court is examined in

Cheng, William, Leana Golubchik, and David Kay. "Total Recall: Are Privacy Changes Inevitable?: A Position Paper." *Proceedings of the First ACM Workshop on Continuous Archival and Retrieval of Personal Experiences* (CARPE '04), New York, October 15, 2004, 86–92.

An overview of Privacy, Ownership, Search, Cryptography, and many of the critical aspects of our lives in Cyberspace:

Abelson, Hal, Ken Ledeen, and Harry Lewis. 2008. *Blown to Bits: Your Life, Liberty, and Happiness After the Digital Explosion*. Boston: Addison Wesley.

We only touched on a little of what Steve Mann has done in wearable computers and mediated reality. His work runs the gamut from technical issues to legal issues to art.

Mann, S. "'Sousveillance': Inverse Surveillance in Multimedia Imaging." *Proceedings of ACM Multimedia 2004,* New York, October 2004, 620–27.

Mann, Steve. 2001. *Intelligent Image Processing.* New York: John Wiley and Sons.

Mann, Steve, Anurag Sehgal, and James Fung. "Continuous Lifelong Capture of Personal Experience Using Eyetap." *Proceedings of The First ACM Workshop on Continuous Archival and Retrieval of Personal Experiences (CARPE '04),* New York, October 15, 2004, 1–21.

Mann, Steve, with Hal Niedzviecki. 2001. *Cyborg: Digital Destiny and Human Possibility in the Age of the Wearable Computer.* New York: Random House Doubleday.

Mann's "EyeTap" glass can record what he sees—now Canadian filmmaker Rob Spence plans to turn his prosthetic eye into a video camera.

"Anti-Surveillance Filmmaker Plans Eye-Socket Camera." Reuters, March 5, 2009.

http://www.eyeborgblog.com

Hasan M. Elahi is an artist exploring the bounds of surveillance and sousveillance by being a more extreme and consistent life blogger. He captures nearly everything that happens in his waking life in photos with time and location. Hasan's tracking site provides an image of exactly where Hasan is at any time.

Hasan Elahi's Web site. http://elahi.org
Hasan's tracking site. http://www.trackingtransience.net

9. GETTING STARTED

Aimee Baldridge has written a complete and excellent how-to for digitizing everything in your life. She describes how you go about estimating, planning, and digitizing your collection of documents, photos, cassettes, videotapes, vinyl records, et cetera. This two-hundred-page book makes it clear just how daunting digitizing your entire past can be. Our recommendation is to start now by accumulating everything that is born digital and go back in time based on need.

Baldridge, Aimee. 2009. *Organize Your Digital Life: How to Store Your Photographs, Music, Videos, and Personal Documents in a Digital World.* Washington, D.C.: National Geographic.

The end of paper is nowhere in sight. Copiers and computer printers continue to accelerate the growth in the use of paper and printing. The first step of going paperless is not storing or transmitting paper—the convenience of paper as a write-once, high quality, portable display will continue until display technology advances quite a bit more.

Paper digitizers, aka scanners, are used to eliminate paper storage and transmission. An ideal scanner would be small, fast, and low cost; it would be able to handle stacks of one- or two-sided items in arbitrary formats and sizes from business cards and photos to books and blueprints, and to scan black-and-white as well as color at arbitrary resolutions, would have a shredder as its last stage, and would be as easy to use as the shredder. In 2010 it takes a half-dozen scanners plus a shredder to realize the ideal. Office copiers with scanning capability are increasingly approaching the ideal because they are fast and require the high resolution for printing that is inherent with an office copier.

The following scanners are in order of my own personal preference:

- *Small units* that fit nicely at the back of a desk. Usually this will be a feed-through format to save space. The Fujitsu ScanSnap models are very nice, handling both sides of the paper in a single pass, both color and black-and-white, at five to fifteen pages per minute.
- *Digital cameras* suffice as scanner alternatives. A high-resolution digital camera can digitize almost anything. An A-size 8½" x 11" page can be resolved at 150 dpi with 2 megapixels or 300 dpi with 8 megapixels. A copy stand or tripod and proper lighting are essential.
- *All-in-one personal print-scan-copy-fax.* A single device scans and prints

documents and photos. Quality, reliability, and speed are often unimpressive.

- *Personal flatbed* with document feeder. These are typically inexpensive, large, and bulky, with a relatively slow feeder (five to eight pages per minute) yet able to scan almost anything, including 3-D objects. Professional flatbed scanners operate at up to twenty-five double-sided pages per minute.
- *Business card scanner.* Special scanner of business cards. Desktop scanners usually handle this, making them redundant.
- *Photo scanner (slides, negatives, positives, and prints).* Photo scanners require the ability to handle multiple items at high resolution (>1,000 dpi), including slides. Photo scanning services are an excellent alternative to owning a photo scanner.
- *Book-specialized flatbed scanners* cost a few hundred dollars but require setup time per page scanned. At the high end of the spectrum, a fully automatic book scanner sells for tens of thousands of dollars.
- *Large format,* e.g., for blueprints. There are a variety of special scanners for scanning large items such as plans and blueprints. Most people will not have very many of these to scan, and will better off using a service to scan large items. An alternative is stitching multiple scans together using image-editing software.

Photo- and slide-scanning services are everywhere, from drugstores to photo stores. The Web lists thousands of them. *Document scanning services* also abound. Some scanning services will also come to your facility to do the scanning. VendorSeek and RecordNation will get you several quotes from different vendors for scanning your documents.

http://vendorseek.com
http://recordnation.com

If you are handling scanned documents, PDF is likely the format you will settle on, and PDF tools for creation, OCR, and editing will be essential.

Adobe Acrobat Web site. https://www.acrobat.com

Nuance Web site. www.nuance.com
CutePDF Web site. http://www.cutepdf.com

Tools for recording family and small business finances:

Microsoft Money. http://www.microsoft.com/MONEY/default.mspx
Quicken. http://www.quicken.com
Mind Your Own Business. http://myob.com

Tools for music ripping and management:

Apple iTunes Web site. http://www.itunes.com
Microsoft Zune Web site. http:// www.zune.net
WinAmp Web site. http://www.winamp.com

Photo editing and management software:

Adobe Elements Web site. http://www.adobelements.com
Adobe PhotoShop Web site. http://www.photoshop.com
Apple iLife Web site. http://www.apple.com/iLife
Picassa Web site. http://picassa.google.com
Microsoft Live Photo Gallery. http://download.live.com/photogallery

Video editing and management software:

Adobe Premier Web site. http://www.adobe.com/products/premiere
Apple iMovie Web site. http://www.apple.com/ilife/imovie
Microsoft Movie Maker Web site. http://www.microsoft.com/window
    sxpusing/moviemaker/default.mspx
Magix Movie Edit Web site. http://www.magix.com/us/video
Pinnacle Studio Web site. http://www.pinnaclesys.com
Ourpix Web site. http://www.ourpix.com
Movie Story Web site. http://ourpix/dvd-slid-show.html

See the Health section below for devices, especially by Philips, Bian-
caMed, BodyMedia, Oregon Scientific, and Omron. See that section also

for health software, including Google Health, Microsoft HealthVault, and Quicken Health.

Microsoft OneNote uses a notebook metaphor for putting "everything," e.g., handwritten notes, voice comments, e-mail messages, writing, Web pages, photos, et cetera, into a searchable hierarchy of notebooks, sections, and pages.

One Note Web page. http://office.microsoft.com/onenote

Evernote, patterned after OneNote and inspired by MyLifeBits, is a cloud service that includes the use of a cell phone for transferring images, video, and voice into the notebooks where it is stored and recognized for access.

Evernote Web site. http://www.evernote.com

Livescribe's Pulse Smartpen is a pen with audio recorder enabling handwritten notes and text input. When handwritten notes become part of computer documents, the voice recordings can be retrieved by pointing to the text. The pen has numerous applications; one can use it as a calculator, recorder, and to record meetings. IOgear has the competing Digital Scribe pen.

Livescribe Web site. http://www.livescribe.com
IOgear Digital Scribe pen. http://www.iogear.com/product/GPEN100C

IBM Pensieve:

Aizenbud-Reshef, Neta, et al. "Pensieve: Augmenting Human Memory." Conference on Human Factors in Computing Systems. Florence, Italy, 2008.

Oregon Scientific and Hammacher Schlemmer have body-mounted cameras. Hammacher Schlemmer even has a diving mask with built-in camera.

Oregon Scientific. http://www2.oregonscientific.com
Hammacher Schlemmer. http://www.hammacher.com

Audio recording is easily accomplished with small recorders from Sony, Philips, Olympus, and others.

Martin, James A. 2008. "The Best Digital Voice Recorder." *PC World* (April 23).
Digital Voice Recorders, Consumer search report. http://www.consum ersearch.com/digital-voice-recorders

For GPS logging, I am currently using a Semsons i-Blue 747. Garmin, Magellan, and others also make devices.

Semsons Web site. http://www.semsons.com
Garmin Web site. http://www.garmin.com
Magellan Web site. http://www.magellangps.com

Geo-tagging:

Microsoft ProPhoto Tools Web site. http://www.microsoft.com/pro photodownloads/tools.aspx
HoudahGeo Web site. http://www.houdah.com/houdahGeo/

Yahoo Research has some excellent people working on geo-tagging. Their ZoneTag lets you upload you photos to Flickr with location tags.

ZoneTag Web site. http://zonetag.research.yahoo.com

Eisenberg describes cloud storage and sharing services of files, including Syncplicity and Dropbox.

Eisenberg, Anna. 2009. "Digital Storage Options for Workers on the Go." *The New York Times* (January 17).
Dropbox Web site. http://www.getdropbox.com
Syncplicity Web site. http://www.syncplicity.com

Tools for home storage and media serving:

Microsoft Windows Home Server Web page. http://www.microsoft.com/ windows/products/winfamily/windowshomeserver/default.mspx

Apple's Time Machine Web page. http://www.apple.com/macosx/ features/timemachine.html

Apple's Time Capsule Web page. http://www.apple.com/timecapsule

Microsoft Windows Media Center Web site. http://www.microsoft.com/ windows/windows-vista/features/media-center.aspx

Apple TV. http://www.apple.com/appletv

Photo frames increasingly offer the ability to distribute photos through-out your home and include the ability to display photos, slide shows, video, and music.

Jacobowitz, P. J., and Zach Honig. 2008. "How to Buy a Digital Photo Frame." PCMag.com (December 12). http://www.pcmag.com/article2/ 0,2817,2300977,00.asp

Digital Photo Frames, cnet reviews. http://reviews.cnet.com/digital-picture-frames

Family Genealogy databases, family-tree making, and time lines are avail-able on the Web, e.g., FamilyTreeMaker.

FamilyTreeMaker Web site. http://www.FamilyTreeMaker.com

The top levels of Gordon Bell's folder hierarchy, circa 2009. While there is no "right" organizing principle, it is a big mistake to not have an orga-nizing principle, or to apply one inconsistently. The full Bell folder hier-archy contains more than 2,000 folders, holding more than 130,000 items. The top division is between active and archive items.

    1) My Documents *holding active content*
        a) Administrative and Systems
        b) CyberAll aka MyLifeBits Research
            i) Papers, patents

      ii) Presentations

      iii) Project plans

      iv) Hardware (eBook, SenseCam, et cetera)

      v) Conferences and other papers

      vi) Classification, facets, metadata

      vii) Database

      viii) This book

   c) Media-in-the-home research

   d) Telepresence research

   e) Systems of all kinds, chips to supers

   f) Other active company and organizational (companies, government, schools)

   g) GB in process books and papers

   h) GB family financial and legal

      i) CYxx or FYxx: Yearly bank, brokerage, expense, receipt, detailed tax, statements

      ii) Investments

      iii) Money and historical transactions

      iv) Property

         (1) Real estate (folder per property)

         (2) A/V, cameras, electronics, phones

         (3) . . .

      v) Start-ups (100+ company folders with business plan, history, ownership and official signed docs, stock certificate TIFFs)

      vi) Tax for all years

      vii) Wills, trusts, and family legal

      viii) Family member(s) financial

   i) GB personal and family members

      i) X's (for all X family members), including: Certificates (birth, . . . passport, et cetera)

      ii) Health X (for all X family members)

      iii) Society memberships

      iv) Vitae

2) My Documents–Archive *holding archival content*

   a) A Big Shoebox *for most personal photos and video snippets*

      i) Family *by year and albums*

      ii) Person(s) x

      iii) Places

      iv) Trips

      v) SenseCam (o(1000) sequences of birthdays, celebrations, conferences, show, walks, et cetera)

      vi) Things

      vii) Food *dishes, restaurants, et cetera*

      viii) Ephemera, memorabilia

      ix) Animals

      x) Underwater photos

b) Profession-specific photos *for artifacts, computers, robots, people*

c) My bio including articles, events, interviews, patents

d) My publications *papers and reports*

e) My talks and presentations

f) Other publications *papers and reports*

g) People, references, recommendations, vitae

h) Archived company and organizational folders (X)

      i) Digital Equipment Corp. . . .

      ii) NSF

i) Archived calendars and correspondence (t)

j) Archived files (e.g., DEC WPS, e-mail)

3) My Books *books authored, books scanned*

4) My Voice Conversations and Notes (telephone conversations are held in MyLifeBits database)

5) My Media, i.e., song collections from ripped CDs

6) My Videos including c. 1950s 8mm movies and lectures

Psychologists have identified "lifetime periods" as an important way that autobiographical memories work. Lifetime periods are thematic and include work or jobs, educational institutions, and relationships that exist over an extended period of time. These lifetime periods are an important part of my hierarchy, above.

Conway, M. A. 2005. "Memory and the Self." *Journal of Memory and Language* 53:594–628.

Conway, M. A., and C. W. Pleydell-Pearce. 2000. "The Construction of Autobiographical Memories in the Self-Memory System." *Psychological Review* 107, no. 2:261–68.

McNeely, I. F., and L. Wolverton. 2008. *Reinventing Knowledge: From Alexandria to the Internet.* New York: W. W. Norton and Company.

Hierarchies can be useful, but are sometimes too constricting. Many times you want to just access items by some attribute. For instance, you can probably access your e-mail by the attributes of date, subject, and sender, without designating that each of them occupies a particular level of a hierarchy. You can also see this when you shop at, say, Amazon, where you can sift cameras by brand and number of pixels. In the world of librarians, using attributes to organize things is called "faceted classification." The Flamenco Search Interface project from the UC Berkeley School of Information has some good demos on faceted organization.

Flamenco Web page. http://flamenco.berkeley.edu

Hearst, Marti A. "UIs for Faceted Navigation: Recent Advances and Remaining Open Problems, in the Workshop on Computer Interaction and Information Retrieval," HCIR 2008, Redmond, Washington, October 2008.

For those of you interested in trying a start-up:

Bell, C. Gordon, and John E. McNamara. 1991. *High-Tech Ventures: The Guide for Entrepreneurial Success.* New York: Perseus Book Publishing.

Nesheim, John L. 2000. *High Tech Start Up, Revised and Updated: The Complete Handbook For Creating Successful New High Tech Companies.* New York: The Free Press.

IO. THE FUTURE

Nathan Myhrvold said, "On the hardware side, I'm pretty confident there'll be another twenty years at least, which is another factor of a million. A factor of a million reduces a year into thirty seconds. Twenty years from now, a computer will do in thirty seconds what one of today's computers would take a year to do. So, for particularly big computational problems there's no point in starting. You should wait, and then do it all in thirty seconds twenty years from now! That is the hardware side. The growth of software is certain, because it's only limited by human imagination."

Brand, Stewart. 1995. "The Physicist." *Wired* (September).

Bell's law:

Bell, G. 2008. "Bell's Law for the Birth and Death of Computer Classes." *Communications of the ACM* 51 (1) (January): 86–94.

Sensors:

Warneke, Brett, Matt Last, Brian Liebowitz, and Kristofer S. J. Pister. 2001. "Smart Dust: Communicating with a Cubic-Millimeter." *Computer* 34: 44–51.
http://robotics.eecs.berkeley.edu/~pister/SmartDust/
http://ceng.usc.edu/~anrg/SensorNetBib.html
Zhao, Feng, and Leonidas Guibas. 2004. *Wireless Sensor Networks: An Information Processing Approach*. San Francisco: Morgan Kaufmann.
ACM Transactions on Sensor Networks. http://tosn.acm.org

Sensors will be everywhere, and so will computational power. The ubiquitous and pervasive computing research communities are driving this forward. There are journals and magazines, including *IEEE Pervasive Computing*, Springer's *Personal and Ubiquitous Computing*, and Elsevier's *Pervasive and Mobile Computing* journal.

Weiser, Mark. 1991. "The computer for the 21st century." *Scientific American* 265, no. 3 (September): 94–104.
Ubiquitous computing conference Web site. http://ubicomp.org
Pervasive conference Web site. http://www.pervasive-conference.org

Pattie Maes's lab at MIT, which previously did the Remembrance Agent work, has recently unveiled a system called "sixth sense" that includes a wearable projector and camera. Any surface becomes a possible computer display, and you control the system using hand gestures (your hands are tracked by the camera). A hand gesture takes a snapshot. A virtual keyboard can be displayed to type on. Bar codes can be scanned while shopping and extra information about products projected on them. When you read the newspaper, extra video footage for a story can be played. When you read a book, annotations and extra information can be projected into the margins. Sixth Sense is a great illustration of the technological climate that Total Recall will live in: constant recording possible with a wearable camera, and your e-memory ready to consult and enjoy at any moment on any surface.

Sixth Sense demo video. http://www.ted.com/talks/pattie_maes_demos_the_sixth_sense.html

Sixth Sense Web page. http://www.pranavmistry.com/projects/sixth-sense.html

Mistry, P., P. Maes, and L. Chang. "WUW—Wear Ur World—A Wearable Gestural Interface." To appear in the CHI '09 *Extended Abstracts on Human Factors in Computing Systems,* Boston, Massachusetts, 2009.

Storage trends: See Chapter 1.

Unified communications: "Forrester Research said recently that the unified communications (UC) market in North America, Europe, and Asia Pacific will reach $14.5bn (£10.5bn) in 2015."

Bell, Gordon, and Jim Gemmell. 2002. "A Call for the Home Media Network. *Communications of the ACM* 45, no. 7 (July): 71–75. Association for Computing Machinery, Inc.

Montalbano, Elizabeth. 2008. "IBM Pledges $1 Billion to Unified Communications." *PC World* (March 11).

O'Reilly, Paul. 2009. "Managing Unified Communications Performance." CRN (March 9).

Semantic Web:

Berners-Lee, T., and J. Hendler. 2001. "Scientific Publishing on the Semantic Web." *Nature* (26 April).

W3C Semantic Web Frequently Asked Questions. http://www.w3.org/RDF/FAQ

British Library Digital Lives Project and conference:

Digital Lives Research Project Web page. http://www.bl.uk/digital-lives
First Digital Lives Research Conference: Personal Digital Archives for the 21st Century. British Library, St. Pancras, London, February 9–11, 2009.

Randy Hahn helped us craft the story about him. The details are fictitious, but the scenario is completely realistic.

ACKNOWLEDGMENTS

We have many to thank for helping make this book a reality. Roger Lueder was our key collaborator throughout the MyLifeBits project. It was the vision of our agent, James Levine, that spurred us on toward a much broader audience. Our editor, Stephen Morrow, guided us out of the mire of the scientific writing style. We are very grateful to Bill Gates for his excellent foreword, and for his technological vision that was a key inspiration for this work. Others who provided invaluable help with the manuscript include Sheridan Forbes, Ray Ozzie, Dave Gemmell, David Rollo, Randy Hahn, and Michael Hahn.

Thanks also to others who helped with MyLifeBits and CARPE: Aleks Aris, Josh Blumenstock, Mary Czerwinski, Steve Drucker, Jonathan Fay, Steve Hodges, Ron Logan, Kenji Mase, Brian Meyers, Tripp Millican, Evan Salomon, Hari Sundaram, Kentaro Toyama, Curtis Wong, Zhe Wang, and Ken Wood. The arrival of Lyndsay Williams's SenseCam in 2003 added another dimension to the project.

Gordon would like to thank his family and friends for their encouragement and especially the enrichment of MyLifeBits. Vicki Rozycki digitized the thousands of items constituting MyLifeBits that enables him to have a real, living, working, essential e-memory.

Jim would like to thank Elizabeth, Sam, Naomi, Judah, Levi, Miriam, and Sara for their support through the hectic writing schedule.

Finally, we want to thank Microsoft, Microsoft Research, and especially Jim Gray. We had the privilege of working for Jim Gray for more than ten years in his Bay Area Research Center in San Francisco. For his Turing Award lecture, Jim outlined some key future research goals, one of them being: "Personal Memex: Record everything a person sees and hears, and quickly retrieve any item on request." Without Jim's support, there would have been no MyLifeBits, and we would not have written this book.

INDEX